Springer Optimization and Its Applications

Volume 145

Aims and Scope
Optimization has been expanding in all directions at an astonishing rate during the last few decades. New algorithmic and theoretical techniques have been developed, the diffusion into other disciplines has proceeded at a rapid pace, and our knowledge of all aspects of the field has grown even more profound. At the same time, one of the most striking trends in optimization is the constantly increasing emphasis on the interdisciplinary nature of the field. Optimization has been a basic tool in all areas of applied mathematics, engineering, medicine, economics and other sciences.

The series *Springer Optimization and Its Applications* publishes undergraduate and graduate textbooks, monographs and state-of-the-art expository works that focus on algorithms for solving optimization problems and also study applications involving such problems. Some of the topics covered include nonlinear optimization (convex and nonconvex), network flow problems, stochastic optimization, optimal control, discrete optimization, multi-objective programming, description of software packages, approximation techniques and heuristic approaches.

More information about this series at http://www.springer.com/series/7393

Ioannis C. Demetriou • Panos M. Pardalos
Editors

Approximation and Optimization

Algorithms, Complexity and Applications

 Springer

Editors
Ioannis C. Demetriou
Department of Economics
University of Athens
Athens, Greece

Panos M. Pardalos (iD)
Department of Industrial & Systems
Engineering
University of Florida
Gainesville, FL, USA

ISSN 1931-6828   ISSN 1931-6836   (electronic)
Springer Optimization and Its Applications
ISBN 978-3-030-12769-5   ISBN 978-3-030-12767-1   (eBook)
https://doi.org/10.1007/978-3-030-12767-1

This Springer imprint is published by the registered company Springer Nature Switzerland AG.
The registered company address is: Gewerbestrasse 11, 6330 Cham, Switzerland

Preface

This volume contains most of the invited papers that were presented at the Conference on Approximation and Optimization, held in Athens, Greece, on 29–30 June 2017.

The occasion being the 180 years celebration of the National and Kapodistrian University of Athens, the conference covered research issues in approximation and optimization by focusing on the development of algorithms, the study of their complexity, and relevant applications.

The individual papers have been written by leading experts and active researchers in their subjects. They are a mix of expository articles, surveys of new work, and applications. The topics have been drawn from approximation to discrete noisy data, data-dependent approximation, evolutionary optimization, machine learning, non-linearly constrained optimization, optimal design of smart composites, optimization of multiband electric filters, portfolio selection, tax evasion as an optimal control problem, and the no-free-lunch theorem.

The book by content, expertise, and application areas will be useful to academics, researchers, industry experts, data science practitioners, business analysts, social sciences investigators, and graduate students.

Support for this conference came from the M.Sc. Program in Business Administration, Analytics, and Information Systems of the University of Athens and the Bank of Greece. We are grateful to our colleague Professor Yannis Stournaras, Governor of the Bank of Greece. Their support was crucial to the academic excellence of the program, to the participation from a wide range of countries, to the social activities, and to the publication of these proceedings. The conference received valuable assistance from the National and Kapodistrian University of Athens through Mrs. Katerina Skoura (Head of Administration) in the organization of the conference and staff in the Department of Economics. To all, we express our sincere thanks.

The authors of the papers deserve our editorial thanks for producing the papers so well and so promptly. Thanks are also due to the referees who were generous with their time and effort. And thanks also to Razia Amzad and the staff of Springer for their help with publishing this book. It is a pleasure to acknowledge all of these contributions.

Athens, Greece   Ioannis C. Demetriou
Gainesville, FL, USA   Panos M. Pardalos

Contents

Contributors

Stavros P. Adam Department of Informatics and Telecommunications, University of Ioannina, Arta, Greece

Computational Intelligence Laboratory – CILab, Department of Mathematics, University of Patras, Patras, Greece

Stamatios-Aggelos N. Alexandropoulos Computational Intelligence Laboratory – CILab, Department of Mathematics, University of Patras, Patras, Greece

Christos K. Aridas Computational Intelligence Laboratory – CILab, Department of Mathematics, University of Patras, Patras, Greece

Andrei Bogatyrëv Institute for Numerical Mathematics, Russian Academy of Sciences, Moscow, Russia

Coralia Cartis Mathematical Institute, Oxford University, Oxford, UK

Michael P. Cullinan Maryvale Institute, Birmingham, UK

Ioannis C. Demetriou Department of Economics, University of Athens, Athens, Greece

Ding-Zhu Du Department of Computer Science, University of Texas at Dallas, Richardson, TX, USA

Georgia Foutsitzi Department of Informatics and Telecommunication, University of Ioannina, Preveza, Greece

Nicholas I. M. Gould Numerical Analysis Group, Rutherford Appleton Laboratory, Chilton, UK

Dimitrios Hristu-Varsakelis Computational Methods and Operations Research Laboratory, Department of Applied Informatics, University of Macedonia, Thessaloniki, Greece

Valery A. Kalyagin Laboratory of Algorithms and Technologies for Network Analysis, National Research University Higher School of Economics, Nizhny Novgorod, Russia

Sotiris B. Kotsiantis Computational Intelligence Laboratory – CILab, Department of Mathematics, University of Patras, Patras, Greece

Yi Li Department of Computer Science, University of Texas at Dallas, Richardson, TX, USA

Paraskevi Papadopoulou Computational Methods and Operations Research Laboratory, Department of Applied Informatics, University of Macedonia, Thessaloniki, Greece

Panos M. Pardalos Department of Industrial & Systems Engineering, University of Florida, Gainesville, FL, USA

Sergey V. Slashchinin Laboratory of Algorithms and Technologies for Network Analysis, National Research University Higher School of Economics, Nizhny Novgorod, Russia

Georgios E. Stavroulakis School of Production Engineering and Management, Technical University of Crete, Institute of Computational Mechanics and Optimization, Chania, Greece

Georgios K. Tairidis School of Production Engineering and Management, Technical University of Crete, Institute of Computational Mechanics and Optimization, Chania, Greece

Philippe L. Toint Namur Center for Complex Systems (naXys) and Department of Mathematics, University of Namur, Namur, Belgium

Michael N. Vrahatis Computational Intelligence Laboratory – CILab, Department of Mathematics, University of Patras, Patras, Greece

Weili Wu Department of Computer Science, University of Texas at Dallas, Richardson, TX, USA

Introduction

Ioannis C. Demetriou and Panos M. Pardalos (iD)

Abstract A brief survey is given to the papers of this volume that explore various aspects of approximation and optimization.

1 Survey

Approximation and optimization form important disciplines within mathematics that have a significant contribution to knowledge, applied knowledge, and computing. Many optimization problems occur naturally and many problems require approximation techniques to be solved. There is an explosion of methods and applications of these disciplines in recent times throughout science, engineering, technology, medicine, and social sciences.

The papers of this volume explore various aspects of approximation and optimization. Some present summaries of the state of the art in a particular subject and some others present new research results. A brief survey of the papers is given below.

Coralia Cartis, Nicholas Gould, and Philippe L. Toint in their paper *Evaluation Complexity Bounds for Smooth Constrained Nonlinear Optimization using Scaled KKT Conditions and High-order Models* consider evaluation complexity for solving convexly constrained optimization. They show that the complexity bound of $O(\epsilon^{-3/2})$ of their previous work for computing an ϵ-approximate first-order critical point can be obtained under significantly weaker assumptions. Further, they generalize this result to the case where high-order derivatives are available, the order being p say, resulting in a bound of $O(\epsilon^{-(p+1)/p})$ evaluations. Then, defining

I. C. Demetriou (✉)
Department of Economics, University of Athens, Athens, Greece
e-mail: demetri@econ.uoa.gr

P. M. Pardalos
Department of Industrial & Systems Engineering, University of Florida, Gainesville, FL, USA
e-mail: pardalos@ufl.edu

© Springer Nature Switzerland AG 2019
I. C. Demetriou, P. M. Pardalos (eds.), *Approximation and Optimization*,
Springer Optimization and Its Applications 145,
https://doi.org/10.1007/978-3-030-12767-1_1

ϵ_P and ϵ_D to be the primal and dual accuracy thresholds, they also show that the bound of $O(\epsilon_P^{-1/2}\epsilon_D^{-3/2})$ evaluations for the general nonconvex case involving both equality and inequality constraints can be generalized to yield a bound of $O(\epsilon_P^{-1/p}\epsilon_D^{-(p+1)/p})$ evaluations under similarly weaker assumptions.

Data-dependent approximation is a new approach for the study of nonsubmodular optimization problems. It has attracted a lot of research especially in the area of social computing, where nonsubmodular combinatorial optimization problems have been recently formulated. Weili Wu, Yi Li, Panos Pardalos, and Ding-Zhu Du in their paper *Data-Dependent Approximation in Social Computing* present some theoretical results and discuss on related problems open to research.

Machine learning algorithms build efficient descriptive or predictive models to uncover relationships in the data. Multi-objective evolutionary optimization assists machine learning algorithms to optimize their hyperparameters, usually under conflicting performance objectives, and selects the best model for a prescribed task. Stamatios-Aggelos Alexandropoulos, Christos Aridas, Sotiris Kotsiantis, and Michael Vrahatis in their paper *Multi-Objective Evolutionary Optimization Algorithms for Machine Learning, a Recent Survey* consider relevant approaches for four major data mining and machine learning tasks, namely data preprocessing, classification, clustering, and association rules.

Optimization search and supervised learning are the areas that have benefited more from the concept of the no free lunch (NFL) theorem. In its rapid growth, NFL has provided new research results, which are also important in other scientific areas where the successful exploration of a search space is an essential task. Stavros Adam, Stamatios-Aggelos Alexandropoulos, Panos Pardalos, and Michael Vrahatis in their paper *No Free Lunch Theorem, a Review* survey research results in this field, reveal main issues, and expose particular points that are helpful in understanding the hypotheses, the restrictions, or even the inability of applying NFLs.

Revd. Michael Cullinan in his paper *Piecewise Convex-Concave Approximation in the Minimax Norm* presents an efficient algorithm for constructing an approximant to noisy data in order to obtain piecewise convexity/concavity with respect to the least uniform change to n data. Specifically, if q sign changes are allowed in the second order consecutive divided differences of the components of the approximant, then, own to the fact that the set of optimal vectors is connected, the least maximum change to the data is computed in only $O(qn \log n)$ operations, a remarkable result indeed. The author develops optimization techniques which obtain the solution by adjustments that depend on local information, so they avoid the disadvantage of the existence of purely local minima.

Ioannis Demetriou in his paper *A Decomposition Theorem for the Least Squares Piecewise Monotonic Data Approximation Problem* considers the least squares change to n univariate data subject to the condition that the first differences of the estimated values have at most q sign changes. The situation compared to the one in the previous paragraph where the objective function is the minimax norm is quite different, because in the least squares case the set of local minima is composed of discrete points. Hence any algorithm that uses local information will stop at a local minimum. Here difficulties are caused by the enormous number of isolated local solutions of the optimization calculation that can occur in practice, namely $O(n^q)$. A theorem is stated that decomposes the problem into least squares monotonic approximation (case $q = 0$) problems to disjoint sets of adjacent data. The decomposition allows the development of a dynamic programming procedure that provides a highly efficient calculation of the solution in only $O(n^2 + qn \log n)$ operations. The solution to the problem is known by previous work of the author, but a proof is presented that provides necessary and sufficient conditions in a unified theorem.

The best uniform rational approximation of the sign function on two intervals was explicitly found by Zolotarëv in 1877, while half a century later this idea entered technology by Cauer's invention of low- and high-pass electrical filters. Andrei Bogatyrëv in his paper *Recent Progress in Optimization of Multiband Electrical Filters* discusses on a recently developed approach for the solution of the optimization problem that arises in the synthesis of multi-band, analogue, digital or microwave, electrical filters, based on techniques from algebraic geometry and generalizations of the Zolotarëv fraction.

Valery Kalyagin and Sergey Slashchinin in their paper *Impact of Error in Parameter Estimations on Large Scale Portfolio Optimization* examine how estimation error for means and covariance matrix of stock returns may affect the results of selected portfolios. They conducted different experiments using both real data from different stock markets and generated samples in order to compare the out-of-sample performance of the estimators and the influence of the estimation error on the portfolio selection. A new surprising phenomenon observed for large-scale portfolio optimization is that the efficiency of the obtained optimal portfolio is biased with respect to the true optimal portfolio.

The main concept of the paper *Optimal Design of Smart Composites* by Georgios Tairidis, Georgia Foutsitzi, and Georgios Stavroulakis stimulates research on the design, optimization, and control issues on smart structures. Optimal design problems related to smart composites are investigated. First, the mechanical properties of a smart composite can be tailored to meet required specifications. Beyond classical shape and layout optimization related to the layers of a composite, pointwise optimization leading to functionally graded composites or even topology optimization can be applied. Furthermore, some basic techniques regarding soft control based on fuzzy and neuro-fuzzy strategies are presented, along with optimization options and methods which can be used for the fine-tuning of the parameters of the system.

Motivated by the persistent phenomenon of tax evasion and the challenge of tax collection during economic crises, Paraskevi Papadopoulou and Dimitri Hristu-Varsakelis in their paper *Tax Evasion as an Optimal Solution to a Partially Observable Markov Decision Process* explore the behavior of a risk-neutral self-interested firm that may engage in tax evasion to maximize its profits. The firm evolves in a tax system which includes many of standard features such as audits, penalties, and occasional tax amnesties, and may be uncertain as to its tax status. They show that the dynamics of the firm can be expressed via a partially observable Markov decision process, use this model to compute the optimal behavior of the firm, and investigate the effect of "leaks" or "pre-announcements" of any tax amnesties on tax revenues. They also compute the effect on firm behavior of any extensions of the statute of limitations within which the firm's tax filings can be audited, and show that such extensions can be a significant deterrent against tax evasion.

Evaluation Complexity Bounds for Smooth Constrained Nonlinear Optimization Using Scaled KKT Conditions and High-Order Models

Coralia Cartis, Nicholas I. M. Gould, and Philippe L. Toint

Abstract Evaluation complexity for convexly constrained optimization is considered and it is shown first that the complexity bound of $O(\epsilon^{-3/2})$ proved by Cartis et al. (IMA J Numer Anal 32:1662–1695, 2012) for computing an ϵ-approximate first-order critical point can be obtained under significantly weaker assumptions. Moreover, the result is generalized to the case where high-order derivatives are used, resulting in a bound of $O(\epsilon^{-(p+1)/p})$ evaluations whenever derivatives of order p are available. It is also shown that the bound of $O(\epsilon_{\rm P}^{-1/2}\epsilon_{\rm D}^{-3/2})$ evaluations ($\epsilon_{\rm P}$ and $\epsilon_{\rm D}$ being primal and dual accuracy thresholds) suggested by Cartis et al. (SIAM J. Numer. Anal. 53:836–851, 2015) for the general nonconvex case involving both equality and inequality constraints can be generalized to yield a bound of $O(\epsilon_{\rm P}^{-1/p}\epsilon_{\rm D}^{-(p+1)/p})$ evaluations under similarly weakened assumptions.

1 Introduction

In [4] and [7], we examined the worst-case evaluation complexity of finding an ϵ-approximate first-order critical point for smooth nonlinear (possibly nonconvex) optimization problems for methods using both first and second derivatives of the objective function. The case where constraints are defined by a convex set was

C. Cartis (✉)
Mathematical Institute, Oxford University, Oxford, UK
e-mail: coralia.cartis@maths.ox.ac.uk

N. I. M. Gould
Numerical Analysis Group, Rutherford Appleton Laboratory, Chilton, UK
e-mail: nick.gould@stfc.ac.uk

Ph. L. Toint
Namur Center for Complex Systems (naXys) and Department of Mathematics, University of Namur, Namur, Belgium
e-mail: philippe.toint@unamur.be

© Springer Nature Switzerland AG 2019
I. C. Demetriou, P. M. Pardalos (eds.), *Approximation and Optimization*,
Springer Optimization and Its Applications 145,
https://doi.org/10.1007/978-3-030-12767-1_2

considered in the first of these references, while the general case (with equality and inequality constraints) was discussed in the second.

It was shown in [4] that at most $O(\epsilon^{-3/2})$ evaluations of the objective function and its derivatives are needed to compute such an approximate critical point. This result, which is identical in order to the best known result for the unconstrained case, comes at the price of potentially restrictive technical assumptions: it was assumed that an approximate first-order critical point of a cubic model subject to the problem's constraints can be obtained for the subproblem solution in a uniformly bounded number of descent steps that is independent of ϵ, that all iterates remain in a bounded set and that the gradient of the objective function is also Lipschitz continuous (see [4] for details). The analysis of [7] then built on the result of the convex case by first specializing it to convexly constrained nonlinear least-squares and then using the resulting complexity bound in the context of a two-phase algorithm for the problem involving general constraints. If ϵ_P and ϵ_D are the primal and the dual criticality thresholds, respectively, it was suggested that at most $O(\epsilon_P^{-1/2}\epsilon_D^{-3/2})$ evaluations of the objective function and its derivatives are needed to compute an approximate critical point in that case, where the Karush–Kuhn–Tucker (KKT) conditions are scaled to take the size of the Lagrange multipliers into account. Because the proof of this result is based on the bound obtained for the convex case, it suffers from the same limitations (not to mention an additional constraint on the relative sizes of ϵ_P and ϵ_D, see [7]).

More recently, Birgin et al. [3] provided a new regularization algorithm for the unconstrained problem with two interesting features. The first is that the model decrease condition used for the subproblem solution is weaker than that used previously, and the second is that the use of problem derivatives of order higher than two is allowed, resulting in corresponding reductions in worst-case complexity. In addition, the same authors also analyzed the worst-case evaluation complexity of the general constrained optimization problem in [2] also allowing for high-order derivatives and models in a framework inspired by that of [6, 7]. At variance with the analysis of these latter references, their analysis considers unscaled approximate first-order critical points in the sense that such points satisfy the standard unscaled KKT conditions with accuracy ϵ_P and ϵ_D.

The first purpose of this paper is to explore the potential of the proposals made in [3] to overcome the limitations of [4] and to extend its scope by considering the use of high-order derivatives and models. A second objective is to use the resulting worst-case bounds to establish strengthened evaluation complexity bounds for the general nonlinearly constrained optimization problem in the framework of scaled KKT conditions, thereby improving [7]. The paper is thus organized in two main sections: Section 2 covering the convexly constrained case and Section 3 allowing general nonlinear constraints. The results obtained are finally discussed in Section 4.

2 Convex Constraints

The first problem we wish to solve is formally described as

$$\min_{x \in \mathscr{F}} f(x), \tag{1}$$

where we assume that $f : \mathbb{R}^n \longrightarrow \mathbb{R}$ is p-times continuously differentiable, bounded from below, and has Lipschitz continuous p-th derivatives. For the q-th derivative of a function $h : \mathbb{R}^n \to \mathbb{R}$ to be Lipschitz continuous on the set $\mathscr{S} \subseteq \mathbb{R}^n$, we require that there exists a constant $L_{h,q} \geq 0$ such that, for all $x, y \in \mathscr{S}$,

$$\|\nabla_x^q h(x) - \nabla_x^q h(y)\|_T \leq (q-1)! \, L_{h,q} \|x - y\|,$$

where $\| \cdot \|_T$ is the recursively induced Euclidean norm on the space of q-th order tensors. We also assume that the feasible set $\mathscr{F}$ is closed, convex and non-empty. Note that this formulation covers standard inequality (and linear equality) constrained optimization in its different forms: the set $\mathscr{F}$ may be defined by simple bounds, and both polyhedral and more general convex constraints. We remark though that we are tacitly assuming here that the cost of evaluating constraint functions and their derivatives is negligible.

The algorithm considered in this paper is iterative. Let $T_p(x_k, s)$ be the p-th order Taylor-series approximation to $f(x_k + s)$ at some iterate $x_k \in \mathbb{R}^n$, and define the local regularized model at x_k by

$$m_k(x_k + s) \stackrel{\text{def}}{=} T_p(x_k, s) + \frac{\sigma_k}{p+1} \|s\|^{p+1}, \tag{2}$$

where $\sigma_k > 0$ is the regularization parameter. Note that $m_k(x_k) = T_p(x_k, 0) = f(x_k)$. The approach used in [4] (when $p = 2$) seeks to define a new iterate x_{k+1} from the preceding one by computing an approximate solution of the subproblem

$$\min_{x \in \mathscr{F}} m_k(x_k + s) \tag{3}$$

using a modified version of the Adaptive Regularization with Cubics (ARC) method for unconstrained minimization. By contrast, we now examine the possibility of modifying the ARp algorithm of [3] with the aim of inheriting its interesting features. As in [4], the modification involves a suitable continuous first-order criticality measure for the constrained problem of minimizing a given function $h : \mathbb{R}^n \to \mathbb{R}$ on $\mathscr{F}$. For an arbitrary $x \in \mathscr{F}$, this criticality measure is given by

$$\pi_h(x) \stackrel{\text{def}}{=} \|P_{\mathscr{F}}[x - \nabla_x h(x)] - x\|, \tag{4}$$

where $P_{\mathscr{F}}$ denotes the orthogonal projection onto $\mathscr{F}$ and $\|\cdot\|$ the Euclidean norm. It is known that x is a first-order critical point of problem (1) if and only if $\pi_f(x) = 0$. Also note that $\pi_f(x) = \|\nabla_x h(x)\|$ whenever $\mathscr{F} = \mathbb{R}^n$.

Next we describe our algorithm as the ARpCC algorithm (ARp for Convex Constraints).

Algorithm 2.1: Adaptive Regularization using p-th order models for convex constraints (ARpCC)

A starting point x_{-1}, an initial and a minimal regularization parameter $\sigma_0 \geq \sigma_{\min} > 0$, algorithmic parameters $\theta > 0$, $\gamma_3 \geq \gamma_2 > 1 > \gamma_1 > 0$ and $1 > \eta_2 \geq \eta_1 > 0$, are given, as well as an accuracy threshold $\epsilon \in (0, 1)$. Compute $x_0 = P_{\mathscr{F}}[x_{-1}]$ and evaluate $f(x_0)$ and $\nabla_x f(x_0)$.
For $k = 0, 1, \ldots$, until **termination**, do:

1. Evaluate $\nabla_x f(x_k)$. If

$$\pi_f(x_k) \leq \epsilon, \tag{5}$$

   terminate. Otherwise compute derivatives of f of order 2 to p at x_k.
2. Compute a step s_k by approximately minimizing $m_k(x_k + s)$ over $s \in \mathscr{F}$ so that

$$x_k + s_k \in \mathscr{F}, \tag{6}$$

$$m_k(x_k + s_k) < m_k(x_k) \tag{7}$$

   and

$$\pi_{m_k}(x_k + s_k) \leq \theta \|s_k\|^p. \tag{8}$$

3. Compute $f(x_k + s_k)$ and

$$\rho_k = \frac{f(x_k) - f(x_k + s_k)}{T_p(x_k, 0) - T_p(x_k, s_k)}. \tag{9}$$

   If $\rho_k \geq \eta_1$, set $x_{k+1} = x_k + s_k$. Otherwise set $x_{k+1} = x_k$.
4. Set

$$\sigma_{k+1} \in \begin{cases} [\max(\sigma_{\min}, \gamma_1 \sigma_k) \sigma_k] & \text{if } \rho_k > \eta_2 & \text{[very successful iteration]} \\ [\sigma_k, \gamma_2 \sigma_k] & \text{if } \eta_1 \leq \rho_k \leq \eta_2 & \text{[successful iteration]} \\ [\gamma_2 \sigma_k, \gamma_3 \sigma_k] & \text{otherwise.} & \text{[unsuccessful iteration],} \end{cases} \tag{10}$$

   and go to step 2 if $\rho_k < \eta_1$.

We first state a useful property of the ARpCC algorithm, which ensures that a fixed fraction of the iterations $1, 2, \ldots, k$ must be either successful or very successful.

Lemma 1 ([3, Lem 2.4], [6, Thm 2.2]) *Assume that, for some $\sigma_{\max} > 0$, $\sigma_j \leq \sigma_{\max}$ for all $0 \leq j \leq k$. Then the ARpCC algorithm ensures that*

$$k \leq \kappa_u |\mathscr{S}_k|, \quad where \quad \kappa_u \stackrel{\text{def}}{=} \left(1 + \frac{|\log \gamma_1|}{\log \gamma_2}\right) + \frac{1}{\log \gamma_2} \log\left(\frac{\sigma_{\max}}{\sigma_0}\right), \tag{11}$$

where $\mathscr{S}_k$ is the number of successful and very successful iterations, in the sense of (10), up to iteration k.

We start our worst-case analysis by formalizing our assumptions

AS.1 The objective function f is p times continuously differentiable on an open set containing $\mathscr{F}$.

AS.2 The p-th derivative of f is Lipschitz continuous on $\mathscr{F}$.

AS.3 The feasible set $\mathscr{F}$ is closed, convex and non-empty.

The ARpCC algorithm is required to start from a feasible $x_0 \in \mathscr{F}$, which, together with the fact that the subproblem solution in Step 2 involves minimization over $\mathscr{F}$, leads to AS.3.

We now recall some simple results whose proof can be found in [3] in the context of the original ARp algorithm.

Lemma 2 *Suppose that AS.1–AS.3 hold. Then, for each $k \geq 0$,*

(i)

$$f(x_k + s_k) \leq T_p(x_k, s_k) + \frac{L_{f,p}}{p} \|s_k\|^{p+1} \tag{12}$$

and

$$\|\nabla_x f(x_k + s_k) - \nabla_s T(x_k, s_k)\| \leq L_{f,p} \|s_k\|^p; \tag{13}$$

(ii)

$$T_p(x_k, 0) - T_p(x_k, s_k) \geq \frac{\sigma_k}{p+1} \|s_k\|^{p+1}; \tag{14}$$

(iii)

$$\sigma_k \leq \sigma_{\max} \stackrel{\text{def}}{=} \max\left[\sigma_0, \frac{\gamma_3 L_{f,p}(p+1)}{p(1 - \eta_2)}\right]. \tag{15}$$

Proof See [3] for the proofs of (12) and (13), which crucially depend on AS.1 and AS.2 being valid on the segmentc $[x_k, x_k + s_k]$, i.e.,

$$\|\nabla_x^p f(x_k + \xi s_k) - \nabla_x^p f(x_k)\|_p \leq L_{f,p} \xi \|s_k\| \quad \text{for all} \quad \xi \in [0, 1]. \tag{16}$$

Observe also that (2) and (7) ensure (14). Assume now that

$$\sigma_k \geq \frac{L_{f,p}(p+1)}{p(1-\eta_2)}. \tag{17}$$

Using (12) and (14), we may then deduce that

$$|\rho_k - 1| \leq \frac{|f(x_k + s_k) - T_p(x_k, s_k)|}{|T_p(x_k, 0) - T_p(x_k, s_k)|} \leq \frac{L_{f,p}(p+1)}{p\,\sigma_k} \leq 1 - \eta_2$$

and thus that $\rho_k \geq \eta_2$. Then iteration k is very successful in that $\rho_k \geq \eta_2$ and $\sigma_{k+1} \leq \sigma_k$. As a consequence, the mechanism of the algorithm ensures that (15) holds. $\diamond$

We now prove that, at successful iterations, the step at iteration k must be bounded below by a multiple of the p-th root of the criticality measure at iteration $k + 1$.

Lemma 3 *Suppose that AS.1–AS.3 hold. Then*

$$\|s_k\| \geq \left[\frac{\pi_f(x_{k+1})}{L_{f,p} + \theta + \sigma_{\max}} \right]^{\frac{1}{p}} \quad \text{for all} \quad k \in \mathscr{S}. \tag{18}$$

Proof Since $k \in \mathscr{S}$ and by definition of the trial point, we have that $x_{k+1} = x_k + s_k$. Observe now that (13) and (15) imply that

$$\|\nabla f(x_{k+1}) - \nabla_x m_k(x_{k+1})\| \leq L_{f,p} \|s_k\|^p + \sigma_k \|s_k\|^p \leq (L_{f,p} + \sigma_{\max}) \|s_k\|^p. \tag{19}$$

Combing this bound with the triangle inequality, the contractive nature of the projection and (8), we deduce that

$$\pi_f(x_{k+1}) = \| P_{\mathscr{F}}[x_{k+1} - \nabla_x f(x_{k+1})] - P_{\mathscr{F}}[x_{k+1} - \nabla_x m_k(x_{k+1})]$$

$$+ P_{\mathscr{F}}[x_{k+1} - \nabla_x m_k(x_{k+1})] - x_{k+1} \|$$

$$\leq \| P_{\mathscr{F}}[x_{k+1} - \nabla_x f(x_{k+1})] - P_{\mathscr{F}}[x_{k+1} - \nabla_x m_k(x_{k+1})] \| + \pi_{m_k}(x_{k+1})$$

$$\leq \| \nabla_x f(x_{k+1})] - \nabla_x m_k(x_{k+1}) \| + \pi_{m_k}(x_{k+1})$$

$$\leq (L_{f,p} + \theta + \sigma_{\max}) \|s_k\|^p$$

and (18) follows. $\diamond$

We now consolidate the previous results by deriving a lower bound on the objective function decrease at successful iterations.

Lemma 4 *Suppose that AS.1–AS.3 hold. Then, if iteration k is successful,*

$$f(x_k) - f(x_{k+1}) \geq \frac{1}{\kappa_s^f} \pi_f(x_{k+1})^{\frac{p+1}{p}},$$

where

$$\kappa_s^f \stackrel{\text{def}}{=} \max \left[1, \frac{p+1}{\eta_1 \sigma_{\min}} \left(L_{f,p} + \theta + \sigma_{\max} \right)^{\frac{p+1}{p}} \right]. \tag{20}$$

Proof If iteration k is successful, we have, using (9), (14), (10), (18) and (15) successively, that

$$f(x_k) - f(x_{k+1}) \geq \eta_1 [T_p(x_k, 0) - T_p(x_k, s_k)]$$

$$\geq \frac{\eta_1 \sigma_{\min}}{p+1} \| s_k \|^{p+1}$$

$$\geq \frac{\eta_1 \sigma_{\min}}{(p+1)[L_{f,p} + \theta + \sigma_{\max}]^{\frac{p+1}{p}}} \pi_f(x_{k+1})^{\frac{p+1}{p}}. \diamond$$

It is important to note that the validity of this lemma does not depend on the history of the algorithm, but is only conditional to the smoothness assumption on the objective function holding along the step from x_k to x_{k+1}. We will make use of that observation in Section 3.

Our worst-case evaluation complexity results can now be proved by combining this last result with the fact that $\pi_f(x_k)$ cannot be smaller than ϵ before termination.

Theorem 1 *Suppose that AS.1–AS.3 hold and let f_{low} be a lower bound on f on $\mathcal{F}$. Then, given $\epsilon > 0$, the ARpCC algorithm applied to problem (1) needs at most*

$$\left\lfloor \kappa_s^f \frac{f(x_0) - f_{\text{low}}}{\epsilon^{\frac{p+1}{p}}} \right\rfloor$$

successful iterations (each involving one evaluation of f and its p first derivatives) and at most

$$\kappa_u \left\lfloor \kappa_s^f \frac{f(x_0) - f_{\text{low}}}{\epsilon^{\frac{p+1}{p}}} \right\rfloor$$

iterations in total to produce an iterate x_ϵ such that $\pi_f(x_\epsilon) \leq \epsilon$, where κ_u is given by (11) with $\sigma_{\max}$ defined by (15).

Proof At each successful iteration, we have, using Lemma 4, that

$$f(x_k) - f(x_{k+1}) \geq (\kappa_s^f)^{-1} \pi_f(x_{k+1})^{\frac{p+1}{p}} \geq (\kappa_s^f)^{-1} \epsilon^{\frac{p+1}{p}},$$

where we used the fact that $\pi_f(x_{k+1}) \geq \epsilon$ before termination to deduce the last inequality. Thus we deduce that, as long as termination does not occur,

$$f(x_0) - f(x_{k+1}) = \sum_{j \in \mathcal{S}_k} [f(x_j) - f(x_j + s_j)] \geq \frac{|\mathcal{S}_k|}{\kappa_s^f} \epsilon^{\frac{p+1}{p}},$$

from which the desired bound on the number of successful iterations follows. Lemma 1 is then invoked to compute the upper bound on the total number of iterations. ◇

For what follows, it is very important to note that the Lipschitz continuity of $\nabla_x^q f$ was only used (in Lemma 2) to ensure that (16) holds for all $k \geq 0$.

3 The General Constrained Case

We now consider the general smooth constrained problem in the form

$$\min_{x \in \mathcal{F}} f(x) \quad \text{subject to} \quad c(x) = 0, \tag{21}$$

where $c : \mathbb{R}^n \to \mathbb{R}^m$ is sufficiently smooth and f and $\mathcal{F}$ are as above. Note that this formulation covers the general problem involving both equality and inequality constraints, the latter being handled using slack variables and the inclusion of the associated simple bounds in the definition of $\mathcal{F}$.

Our idea is now to first apply the ARpCC algorithm to the problem

$$\min_{x \in \mathcal{F}} \nu(x) \stackrel{\text{def}}{=} \tfrac{1}{2} \|c(x)\|^2. \tag{22}$$

If an approximately feasible point is found, then we may follow the spirit of [5–7] and [2] and apply the same ARpCC to approximately solve the problem

$$\min_{x \in \mathcal{F}} \mu(x, t_k) \stackrel{\text{def}}{=} \tfrac{1}{2} \|r(x, t_k)\|^2 \stackrel{\text{def}}{=} \tfrac{1}{2} \left\| \begin{pmatrix} c(x) \\ f(x) - t_k \end{pmatrix} \right\|^2 \tag{23}$$

in the set for some monotonically decreasing sequence of "targets" t_k ($k = 1, \ldots$).

Algorithm 3.1: Adaptive Regularization using p-th order models for general constraints (ARpGC)

A constant β defining $\mathscr{C}_\beta$, a starting point x_{-1}, a minimum regularization parameter $\sigma_{\min} > 0$, an initial regularization parameter $\sigma_0 \geq \sigma_{\min}$ are given, as well as a constant $\delta \in (0, 1)$. The primal and dual tolerances $0 < \epsilon_P < 1$ and $0 < \epsilon_D < 1$ are also given.

Phase 1:

Starting from $x_0 = P_{\mathscr{F}}(x_{-1})$, apply the AR$p$CC algorithm to minimize $\frac{1}{2}\|c(x)\|^2$ subject to $x \in \mathscr{F}$ until a point $x_1 \in \mathscr{F}$ is found such that

$$\|c(x_1)\| < \delta\epsilon_P \quad \text{or} \quad \pi_{\frac{1}{2}\|c\|^2}(x_1) \leq \epsilon_D\|c(x_1)\|. \tag{24}$$

If $\|c(x_1)\| \geq \delta\epsilon_P$, terminate with $x_\epsilon = x_1$.

Phase 2:

1. Set $t_1 = f(x_1) - \sqrt{\epsilon_P^2 - \|c(x_1)\|^2}$.
2. For $k = 1, 2, \ldots$, do:

   a. Starting from x_k, apply the ARpCC algorithm to minimize $\mu(x, t_k)$ as a function of $x \in \mathscr{F}$ until an iterate $x_{k+1} \in \mathscr{F}$ is found such that

   $$\|r(x_{k+1}, t_k)\| < \delta\epsilon_P \quad \text{or} \quad f(x_{k+1}) < t_k \quad \text{or} \quad \pi_\mu(x_{k+1}, t_k) \leq \epsilon_D\|r(x_{k+1}, t_k)\| \tag{25}$$

   b. i. If $\|r(x_{k+1}, t_k)\| < \delta\epsilon_P$, define t_{k+1} according to

   $$t_{k+1} = f(x_{k+1}) - \sqrt{\epsilon_P^2 - \|c(x_{k+1})\|^2}. \tag{26}$$

   and terminate with $(x_\epsilon, t_\epsilon) = (x_{k+1}, t_{k+1})$ if $\pi_\mu(x_{k+1}, t_{k+1}) \leq \epsilon_D\|r(x_{k+1}, t_{k+1})\|$.

   ii. If $\|r(x_{k+1}, t_k)\| \geq \delta\epsilon_P$ and $f(x_{k+1}) < t_k$, define t_{k+1} according to

   $$t_{k+1} = 2f(x_{k+1}) - t_k \tag{27}$$

   and terminate with $(x_\epsilon, t_\epsilon) = (x_{k+1}, t_{k+1})$ if $\pi_\mu(x_{k+1}, t_{k+1}) \leq \epsilon_D\|r(x_{k+1}, t_{k+1})\|$.

   iii. If $\|r(x_{k+1}, t_k)\| \geq \delta\epsilon_P$ and $f(x_{k+1}) \geq t_k$, terminate with $(x_\epsilon, t_\epsilon) = (x_{k+1}, t_k)$

Observe that the recomputations of $\pi_\mu(x_{k+1}, t_{k+1})$ in Step 2.(b) do not require re-evaluating $f(x_{k+1})$ or $c(x_{k+1})$ or any of their derivatives.

We now complete our assumptions.

| AS.4 | All derivatives of f of order 0 to p are Lipschitz continuous in $\mathscr{F}$. |

| AS.5 | For each $i = 1, \ldots, m$, the constraint function c_i is p times continuously differentiable on an open set containing $\mathscr{F}$. |

| **AS.6** | All derivatives of order 0 to p of each c_i ($i = 1, \ldots, m$) are Lipschitz continuous in $\mathscr{F}$. |

| **AS.7** | There exists constants $\beta \geq \epsilon_P$ and $f_{\text{low}} \in \mathbb{R}$ such that $f(x) \geq f_{\text{low}}$ for all $x \in \mathscr{C}_\beta \stackrel{\text{def}}{=} \{x \in \mathscr{F} \mid \|c(x)\| \leq \beta\}$. |

Assume, without loss of generality, that all Lipschitz constants implied by AS.4 and AS.6 are bounded above by $L \geq 1$. Also note the problem of finding an ϵ_P-feasible minimizer of $f(x)$ is only meaningful if AS.7 holds.

We first verify that our assumptions are sufficient to imply that $v(x)$ and $\mu(x, t)$ have Lipschitz p-th derivative on all segments $[x_k, x_k + s_k]$ generated by the algorithm, allowing us to exploit the results of Section 2.

Lemma 5 *Assume that AS.3, AS.5 and AS.6 hold. Let the iteration of the ARpCC algorithm applied to problem (22) be indexed by j. Then the "segment" Lipschitz condition (16) holds for $\nabla_x^q v(x)$ holds on every segment $[x_j, x_j + s_j]$ ($j \geq 0$) generated by the ARpCC algorithm during Phase 1 and any $q \in \{1, \ldots, p\}$. If, in addition, AS.1 and AS.4 also hold, then the same conclusion holds for $\nabla_x^q \mu(x, t)$ on every segment $[x_j, x_j + s_j]$ ($j \geq 0$) generated by the ARpCC algorithm during Step 2.(a) of Phase 2 and any $q \in \{1, \ldots, p\}$, the Lipschitz constant in this latter case being independent of t.*

Proof Since

$$\nabla_x^q v(x) = \sum_{i=1}^{m} \left[\sum_{\ell, j > 0, \, \ell + j = q} \alpha_{\ell, j} \nabla_x^\ell c_i(x) \nabla_x^j c_i(x) + c_i(x) \nabla_x^q c_i(x) \right]$$

(where $\{\alpha_{\ell, j}\}$ are suitable positive and finite coefficients), condition (16) is satisfied on the segment $[x_j, x_j + s_j]$ if (i) the derivatives $\{\nabla_x^{\min[\ell, j]} c_i(x)\}_{i=1}^{m}$ are Lipschitz continuous on $[x_j, x_j + s_j]$, (ii) $\{\nabla_x^{\max[\ell, j]} c_i(x)\}_{i=1}^{m}$ are uniformly bounded on $[x_j, x_j + s_j]$, and (iii) we have that

$$\sum_{i=1}^{m} \|c_i(x_j + \xi s_j) \nabla_x^q c_i(x_j + \xi s_j) - c_i(x_j) \nabla_x^q c_i(x_j)\|_q \leq L_1 \xi \|s_j\| \tag{28}$$

for some constant $L_1 > 0$. The first of these conditions is ensured by AS.6, and the second by the observation that AS.6 again implies that $\|\nabla_x^\ell c_i(x)\| \leq L$ for $\ell \in \{1, \ldots, q\}$ (see [11, Lem. 1.2.2, p. 21]). Moreover,

$$\|c_i(x_j + \xi s_j) \nabla_x^q c_i(x_j + \xi s_j) - c_i(x_j) \nabla_x^q c_i(x_j)\|$$
$$\leq |c_i(x_j + \xi s_j) - c_i(x_j)| \cdot \|\nabla_x^q c_i(x_j + \xi s_j)\|_q$$
$$+ |c_i(x_j)| \cdot \|\nabla_x^q c_i(x_j + \xi s_j) - \nabla_x^q c_i(x_j)\|_q,$$

and the first term on the right-hand side is bounded above by $L^2 \xi \|s_j\|$ and the second by $|c_i(x_j)| L \xi \|s_j\|$. Hence (28) holds with

$$L_1 = \sum_{i=1}^{m} \left(L^2 + |c_i(x_j)| L \right) \le mL^2 + m\|c(x_j)\|L \le mL^2 + m\|c(x_0)\|L$$

because the ARpCC algorithm ensures that $\|c(x_j)\| \le \|c(x_0)\|$ for all $j \ge 0$. As a consequence, AS.3, AS.5 and AS.6 guarantee that (16) holds with the Lipschitz constant

$$m \left[\left(\max_{i=1,\dots,m} \alpha_i \right) L^2 + L^2 + \|c(x_0)\|L \right].$$

If we now assume that AS.1 and AS.4 also hold, we may repeat, for $\mu(x, t)$ (with fixed t) the same reasoning as above and obtain that condition (16) holds for each segment $[x_j, x_j + s_j]$ generated by the ARpCC algorithm applied in Step 2.(a) of Phase 2, with Lipschitz constant

$$m \left[\left(\max_{i=1,\dots,m} \alpha_i \right) L^2 + L^2 + \|c(x_{j,0})\|L \right] + \left(\max_{i=1,\dots,m} \alpha_i \right) L^2 + L^2 + |f(x_{j,0}) - t_j|L$$

$$\le (m+1)L^2 \left[1 + \left(\max_{i=1,\dots,m} \alpha_i \right) \right] + 2mL \overset{\text{def}}{=} L_{\mu,p},$$

where we have used (34) and $\epsilon_P \le 1$ to deduce the inequality. Note that this constant is independent of t_j, as requested. $\diamond$

We now start our complexity analysis as such by examining the complexity of Phase 1.

Lemma 6 *Suppose that AS.3, AS.5 and AS.6 hold. Then Phase 1 of the ARpGC algorithm terminates after at most*

$$\left\lfloor \kappa_{cc}^c \|c(x_0)\| \max \left[\epsilon_P^{-1}, \epsilon_P^{-\frac{1}{p}} \epsilon_D^{-\frac{p+1}{p}} \right] \right\rfloor + 1$$

evaluations of c and its derivatives, where $\kappa_{cc}^c \overset{\text{def}}{=} \frac{1}{2} \kappa_u \kappa_s^{\frac{1}{2}\|c\|^2} \delta^{\frac{1}{p}}$ with $\kappa_s^{\frac{1}{2}\|c\|^2}$ being the problem-dependent constant defined in (20) for the function $\frac{1}{2}\|c(x)\|^2$ corresponding to (22).

Proof Let us index the iteration of the ARpCC algorithm applied to problem (22) by j and assume that iteration j is successful and that

$$\|c(x_j)\| \ge \delta \epsilon_P. \tag{29}$$

If $\|c(x_{j+1})\| \le \frac{1}{2}\|c(x_j)\|$, then

$$\|c(x_j)\| - \|c(x_{j+1})\| \ge \frac{1}{2}\|c(x_j)\| \ge \frac{1}{2}\delta\,\epsilon_{\mathrm{P}}. \tag{30}$$

By contrast, if $\|c(x_{j+1})\| > \frac{1}{2}\|c(x_j)\|$, then, using the decreasing nature of the sequence $\{\|c(x_j)\|\}$, Lemma 4 (which is applicable because of Lemma 5) and the second part of (24), we obtain that

$$(\|c(x_j)\| - \|c(x_{j+1})\|)\,\|c(x_j)\| \ge \frac{1}{2}\|c(x_j)\|^2 - \frac{1}{2}\|c(x_{j+1})\|^2$$

$$\ge \left(\kappa_s^{\frac{1}{2}\|c\|^2}\right)^{-1} (\epsilon_{\mathrm{D}}\|c(x_{j+1})\|)^{\frac{p+1}{p}}$$

$$\ge \left(\kappa_s^{\frac{1}{2}\|c\|^2}\right)^{-1} \left(\tfrac{1}{2}\epsilon_{\mathrm{D}}\|c(x_j)\|\right)^{\frac{p+1}{p}}$$

and thus that

$$\|c(x_j)\| - \|c(x_{j+1})\| \ge \left(\kappa_s^{\frac{1}{2}\|c\|^2}\right)^{-1} 2^{-\frac{p+1}{p}} \|c(x_j)\|^{\frac{1}{p}} \epsilon_{\mathrm{D}}^{\frac{p+1}{p}} \ge \frac{1}{2}\left(\kappa_s^{\frac{1}{2}\|c\|^2}\right)^{-1} \delta^{\frac{1}{p}}\epsilon_{\mathrm{P}}^{\frac{1}{p}}\epsilon_{\mathrm{D}}^{\frac{p+1}{p}},$$

where we have used (29) to derive the last inequality. Because of (20), we thus obtain from this last bound and (30) that, for all j,

$$\|c(x_j)\| - \|c(x_{j+1})\| \ge \frac{1}{2}\left(\kappa_s^{\frac{1}{2}\|c\|^2}\right)^{-1} \delta^{\frac{1}{p}} \min\left[\epsilon_{\mathrm{P}}, \epsilon_{\mathrm{P}}^{\frac{1}{p}}\epsilon_{\mathrm{D}}^{\frac{p+1}{p}}\right].$$

As in Theorem 1, we then deduce that the number of successful iterations required for the ARpCC algorithm to produce a point x_1 satisfying (24) is bounded above by

$$\frac{1}{2}\kappa_s^{\frac{1}{2}\|c\|^2}\,\delta^{\frac{1}{p}}\|c(x_0)\|\max\left[\epsilon_{\mathrm{P}}^{-1}, \epsilon_{\mathrm{P}}^{-\frac{1}{p}}\epsilon_{\mathrm{D}}^{-\frac{p+1}{p}}\right].$$

The desired conclusion then follows by using Lemma 1 and adding one for the final evaluation at termination. ◇

Note that an improved complexity bound for convexly constrained least-squares problems, and hence for Phase 1, was given in [8]. In particular, the bound in Lemma 6 improves to

$$\left\lfloor \kappa_{\mathrm{CC-1}}^c \|c(x_0)\|^{\frac{1}{p}} \max\left[\epsilon_{\mathrm{P}}^{-\frac{1}{p}}, \epsilon_{\mathrm{D}}^{-\frac{p+1}{p}}\right] \right\rfloor + 1$$

whenever p is a power of 2. However, we are not aware of how to use the better Phase 1 result to improve the complexity of Phase 2, so we omit including it here in full.

We now partition the Phase 2 outer iterations (before that where termination occurs) into two subsets whose indexes are given by

$$\mathcal{K}_+ \overset{\text{def}}{=} \{k \geq 0 \mid \|r(x_{k+1}, t_k)\| < \delta \epsilon_{\text{P}} \quad \text{and (26) is applied} \} \tag{31}$$

and

$$\mathcal{K}_- \overset{\text{def}}{=} \{k \geq 0 \mid \|r(x_{k+1}, t_k)\| \geq \delta \epsilon_{\text{P}} \quad \text{and (27) is applied} \}. \tag{32}$$

This partition allows us to prove the following technical results.

Lemma 7 *The sequence $\{t_k\}$ is monotonically decreasing. Moreover, in every Phase 2 iteration of the ARpGC algorithm of index $k \geq 1$, we have that*

$$f(x_k) - t_k \geq 0, \tag{33}$$

$$\|r(x_{k+1}, t_{k+1})\| = \epsilon_{\text{P}} \quad \text{for} \quad k \in \mathcal{K}_+, \tag{34}$$

$$\|r(x_{k+1}, t_{k+1})\| = \|r(x_{k+1}, t_k)\| \leq \epsilon_{\text{P}} \quad \text{for} \quad k \in \mathcal{K}_-, \tag{35}$$

$$\|c(x_k)\| \leq \epsilon_{\text{P}} \quad \text{and} \quad f(x_k) - t_k \leq \epsilon_{\text{P}}, \tag{36}$$

$$t_k - t_{k+1} \geq (1 - \delta)\epsilon_{\text{P}} \quad \text{for} \quad k \in \mathcal{K}_+. \tag{37}$$

Finally, at termination of the ARpGC algorithm,

$$\|r(x_\epsilon, t_\epsilon)\| \geq \delta \epsilon_{\text{P}} \quad \text{and} \quad f(x_\epsilon) \geq t_\epsilon \quad \text{and} \quad \pi_\mu(x_\epsilon, t_\epsilon) \leq \epsilon_{\text{D}} \|r(x_\epsilon, t_\epsilon)\|. \tag{38}$$

Proof The inequality (33) follows from (26) for $k - 1 \in \mathcal{K}_+$ and from (27) for $k - 1 \in \mathcal{K}_-$. Equation (34) is also deduced from (26), while (27) implies the equality in (35), the inequality in that statement resulting from the monotonically decreasing nature of $\|r(x, t_k)\|$ during inner iterations in Step 2.(a) of the ARpGC algorithm. The inequalities (36) then follow from (33), (34) and (35). We now prove (37), which only occurs when $\|r(x_{k+1}, t_k)\| \leq \delta \epsilon_{\text{P}}$, that is when

$$(f(x_{k+1}) - t_k)^2 + \|c(x_{k+1})\|^2 \leq \delta^2 \epsilon_{\text{P}}^2. \tag{39}$$

From (26), we then have that

$$t_k - t_{k+1} = -(f(x_{k+1}) - t_k) + \sqrt{\|r(x_k, t_k)\|^2 - \|c(x_{k+1})\|^2}. \tag{40}$$

Now taking into account that the global minimum of the problem

$$\min_{(f,c) \in \mathbb{R}^2} \psi(f, c) \overset{\text{def}}{=} -f + \sqrt{\epsilon_{\text{P}}^2 - c^2} \quad \text{subject to} \quad f^2 + c^2 \leq \omega^2$$

for $\omega \in [0, \epsilon_P]$ is attained at $(f_*, c_*) = (\omega, 0)$ and it is given by $\psi(f_*, c_*) = \epsilon_P - \omega$ (see [7, Lemma 5.2]), we obtain from (39) and (40) (setting $\omega = \delta\epsilon_P$) that

$$t_k - t_{k+1} \geq \epsilon_P - \omega = (1 - \delta)\epsilon_P \quad \text{for} \quad k \in \mathcal{K}_+$$

for $k \in \mathcal{K}_+$, which is (37). Note that, if $k \in \mathcal{K}_-$, then we must have that $t_k > f(x_{k+1})$ and thus (27) ensures that $t_{k+1} < t_k$. This observation and (37) then allow us to conclude that the sequence $\{t_k\}$ is monotonically decreasing.

In order to prove (38), we need to consider, in turn, each of the three possible cases where termination occurs in Step 2.(b). In the first case (i), $\|r(x_{k+1}, t_k)\|$ is small (in the sense that the first inequality in (25) holds) and (26) is then used, implying that (34) holds and that $f(x_{k+1}) > t_{k+1}$. If termination occurs because $\pi(x_{k+1}, t_{k+1}) \leq \epsilon_D\|r(x_{k+1}, t_{k+1})\|$, then (38) clearly holds at (x_{k+1}, t_{k+1}). In the second case (ii), $\|r(x_{k+1}, t_k)\|$ is large (the first inequality in (25) fails), but $f(x_{k+1}) < t_k$, and t_{k+1} is then defined by (27), ensuring that $f(x_{k+1}) > t_{k+1}$ and, because of (35), that $\|r(x_{k+1}, t_{k+1})\|$ is also large. As before (38) holds at (x_{k+1}, t_{k+1}) if termination occurs because $\pi(x_{k+1}, t_{k+1}) \leq \epsilon_D\|r(x_{k+1}, t_{k+1})\|$. The third case (iii) is when $\|r(x_{k+1}, t_k)\|$ is sufficiently large and $f(x_{k+1}) \geq t_k$. But (25) then guarantees that $\pi(x_{k+1}, t_k) \leq \epsilon_D\|r(x_{k+1}, t_k)\|$, and the inequalities (38) are again satisfied at (x_{k+1}, t_k). $\diamond$

Using the results of this lemma allows us to bound the number of outer iterations in $\mathcal{K}_+$.

Lemma 8 *Suppose that AS.7 holds. Then*

$$|\mathcal{K}_+| \leq \frac{f(x_1) - f_{\text{low}} + 1}{1 - \delta} \epsilon_P^{-1}.$$

Proof We first note that (34) and (35) and AS.7 ensure that $x_k \in \mathcal{C}_\beta$ for all $k \geq 0$. The result then immediately follows from AS.7 again and the observation that, from (37), t_k decreases monotonically with a decrease of at least $(1 - \delta)\epsilon_P$ for $k \in \mathcal{K}_+$. $\diamond$

We now state a very useful consequence of Lemmas 5 and 7.

Lemma 9 *Suppose that AS.1 and AS3–AS.6 hold. Then there exists a constant $\sigma_{\mu,\max} > \sigma_{\min}$ such that all regularization parameters arising in the ARpCC algorithm within Step 2.(a) of the ARpGC algorithm are bounded above by $\sigma_{\mu,\max}$.*

Proof AS.1, AS.4–AS.6 and Lemma 5 guarantee the existence of a Lipchitz constant independent of t such that the "segment-wise" Lipschitz condition (16) holds for each segment $[x_{j,\ell}, x_{j,\ell} + s_{j,\ell}]$. The result is then derived by introducing $L_{\mu,p}$ in (15) to obtain $\sigma_{\mu,\max}$. $\diamond$

The main consequence of this result is that we may apply the ARpCC algorithm to the minimization of $\mu(x, t_k)$ in Step 2.(a) of the ARpGC algorithm and use all the properties of the former (as derived in the previous section) using problem constants valid for every possible t_k.

Consider now x_k for $k \in \mathcal{K}_+$ and denote by $x_{k+\ell(k)}$ the next iterate such that $k + \ell(k) \in \mathcal{K}_+$ or the algorithm terminates at $k + \ell(k)$. Two cases are then possible: either a single pass in Step 2.(a) of the ARpGC algorithm is sufficient to obtain $x_{k+\ell(k)}$ ($\ell(k) = 1$) or two or more passes are necessary, with iterations $k+1, \dots, k+\ell(k) - 1$ belonging to $\mathcal{K}_-$. Assume that the iterations of the ARpCC algorithm at Step 2.(a) of the outer iteration j are numbered $(j, 0), (j, 1), \dots, (j, e_j)$ and note that the mechanism of the ARpGC algorithm ensures that iteration (j, e_j) is successful for all j. Now define, for $k \in \mathcal{K}_+$, the index set of all inner iterations necessary to deduce $x_{k+\ell(k)}$ from x_k, that is

$$\mathscr{I}_k \overset{\text{def}}{=} \{(k, 0), \dots, (k, e_k), \dots, (j, 0), \dots, (j, e_j), \\ \dots, (k + \ell(k) - 1, 0), \dots (k + \ell(k) - 1, e_{k+\ell(k)-1})\}, \tag{41}$$

where $k < j < k + \ell(k) - 1$. Observe that, by definitions (31) and (41), the index set of all inner iterations before termination is given by $\cup_{k \in \mathcal{K}_+} \mathscr{I}_k$, and therefore that the number of evaluations of problem's functions required to terminate in Phase 2 is bounded above by

$$\left| \bigcup_{k \in \mathcal{K}_+} \mathscr{I}_k \right| + 1 \leq \left(\frac{f(x_1) - f_{\text{low}} + 1}{1 - \delta} \epsilon_{\text{P}}^{-1} \times \max_{k \in \mathcal{K}_+} |\mathscr{I}_k| \right) + 1, \tag{42}$$

where we added 1 to take the final evaluation into account and where we used Lemma 8 to deduce the inequality. We now invoke the complexity properties of the ARpCC algorithm applied to problem (23) to obtain an upper bound on the cardinality of each $\mathscr{I}_k$.

Lemma 10 *Suppose that AS.1 and AS.3–AS.6 hold. Then, for each $k \in \mathcal{K}_+$ before termination,*

$$|\mathscr{I}_k| \leq (1 - \delta)\kappa_{CC}^\mu \max\left[1, \epsilon_P^{-\frac{p-1}{p}} \epsilon_D^{-\frac{p+1}{p}} \right],$$

where κ_{CC}^μ is independent of ϵ_P and ϵ_D and captures the problem-dependent constants associated with problem (23) for all values of t_k generated by the algorithm.

Proof Observe first that, because of Lemma 9, we may apply the ARpCC algorithm for the minimization of $\mu(x, t_j)$ for each j such that $k \leq j < k + \ell(k)$. Observe that (35) and the mechanism of this algorithm guarantee the decreasing nature of the sequence $\{\|r(x_j, t_j)\|\}_{j=k}^{k+\ell(k)-1}$ and hence of the sequence $\{\|r(x_{j,s}, t_j)\|\}_{(j,s)\in\mathscr{I}_k}$. This reduction starts from the initial value $\|r(x_{k,0}, t_k)\| = \epsilon_P$ and is carried out for all iterations with index in $\mathscr{I}_k$ at worst until it is smaller than $\delta\epsilon_P$ (see the first part of (25)). We may then invoke Lemmas 9 and 4 to deduce that, if $(j, s) \in \mathscr{I}_k$ is the index of a successful inner iteration and as long as the third part of (25) does not hold,

$$(\|r(x_{j,s},t_j)\| - \|r(x_{j,s+1},t_j)\|)\|r(x_{j,s},t_j)\| \geq \tfrac{1}{2}\|r(x_{j,s},t_j)\|^2 - \tfrac{1}{2}\|r(x_{j,s+1},t_j)\|^2$$

$$\geq \left[\kappa_{CC}^{\mu,s}\right]^{-1}(\epsilon_D\|r(x_{j,s+1},t_j)\|)^{\frac{p+1}{p}},$$

(43)

for $0 \leq s < e_j$ and for some constant $\kappa_{CC}^{\mu,s} > 0$ independent of ϵ_P, ϵ_D, s and j, while

$$\tfrac{1}{2}\|r(x_{j,e_j},t_j)\| - \tfrac{1}{2}\|r(x_{j+1,0},t_{j+1})\| = 0.$$

As above, suppose first that $\|r(x_{j,s+1},t_j)\| \leq \tfrac{1}{2}\|r(x_{j,s},t_j)\|$. Then

$$\|r(x_{j,s},t_j)\| - \|r(x_{j,s+1},t_j)\| \geq \tfrac{1}{2}\|r(x_{j,s},t_j)\| \geq \tfrac{1}{2}\delta\epsilon_P \qquad (44)$$

because of the first part of (25). If $\|r(x_{j,s+1},t_j)\| > \tfrac{1}{2}\|r(x_{j,s},t_j)\|$ instead, then (43) implies that

$$\|r(x_{j,s},t_j)\|-\|r(x_{j,s+1},t_j)\| \geq \left[\kappa_{CC}^{\mu,s}\right]^{-1} 2^{-\frac{p+1}{p}}\|r(x_{j,s},t_j)\|^{\frac{1}{p}}\epsilon_D^{\frac{p+1}{p}} \geq \left[\kappa_{CC}^{\mu,s}\right]^{-1} 2^{-\frac{p+1}{p}}\delta^{\frac{1}{p}}\epsilon_P^{\frac{1}{p}}\epsilon_D^{\frac{p+1}{p}}.$$

Combining this bound with (44) gives that

$$\|r(x_{j,s},t_j)\| - \|r(x_{j,s+1},t_j)\| \geq \left[\kappa_{CC}^{\mu,s}\right]^{-1} 2^{-\frac{p+1}{p}}\delta^{\frac{1}{p}}\min\left[\epsilon_P, \epsilon_P^{\frac{1}{p}}\epsilon_D^{\frac{p+1}{p}}\right].$$

As a consequence, the number of successful iterations of the ARpCC algorithm needed to compute $x_{k+\ell(k)}$ from x_k cannot exceed

$$\kappa_{CC}^{\mu,s}\, 2^{\frac{p+1}{p}}\delta^{-\frac{1}{p}}\left[\frac{\epsilon_P - \delta\epsilon_P}{\min\left[\epsilon_P, \epsilon_P^{\frac{1}{p}}\epsilon_D^{\frac{p+1}{p}}\right]}\right] = (1-\delta)\,\kappa_{CC}^{\mu,s}\, 2^{\frac{p+1}{p}}\delta^{-\frac{1}{p}}\max\left[1, \epsilon_P^{\frac{p-1}{p}}\epsilon_D^{-\frac{p+1}{p}}\right].$$

We now use Lemma 9 again and invoke Lemma 1 to account for possible unsuccessful inner iterations, yielding that the total number of successful and unsuccessful iterations of the ARpCC algorithm necessary to deduce $x_{k+\ell(k)}$ from x_k is bounded above by

$$\kappa_u(1-\delta)\, 2^{\frac{p+1}{p}}\delta^{-\frac{1}{p}}\kappa_{CC}^{\mu,s}\max\left[1, \epsilon_P^{\frac{p-1}{p}}\epsilon_D^{-\frac{p+1}{p}}\right] \overset{\text{def}}{=} (1-\delta)\kappa_{CC}^{\mu}\max\left[1, \epsilon_P^{\frac{p-1}{p}}\epsilon_D^{-\frac{p+1}{p}}\right].\ \diamond$$

We now state a useful property of the set $\mathcal{F}$.

Lemma 11 *For arbitrary $x \in \mathscr{F}$, $v \in \mathbb{R}^n$ and $\tau \in \mathbb{R}$ with $\tau \geq 1$,*

$$\| P_{\mathscr{F}}[x + \tau v] - x \| \leq \tau \| P_{\mathscr{F}}[x + v] - x \|.$$

Proof The result follows immediately from [1, Lem.2.3.1] which states that $\| P_{\mathscr{F}}[x + \tau v] - x \| / \tau$ is a monotonically non-increasing function of $\tau > 0$ for any x in a given convex set $\mathscr{F}$. ◇

We finally combine our results in a final theorem stating our evaluation complexity bound for the ARpGC algorithm.

Theorem 2 *Suppose that AS.1 and AS.3–AS.7 hold. Then, for some constants κ_{cc}^c and κ_{cc}^{μ} independent of ϵ_P and ϵ_D, the ARpGC algorithm applied to problem (21) needs at most*

$$\left\lfloor \left(\kappa_{cc}^c \| c(x_0) \| + \kappa_{cc}^{\mu} [f(x_1) - f_{\text{low}} + 1] \right) \max \left[\epsilon_P^{-1}, \epsilon_P^{-\frac{1}{p}} \epsilon_D^{-\frac{p+1}{p}} \right] + 1 \right\rfloor \quad (45)$$

evaluations of f, c and their derivatives up to order p to compute a point x_ϵ such that either

$$\| c(x_\epsilon) \| > \delta \epsilon_P \quad \text{and} \quad \pi_{\frac{1}{2} \| c \|^2}(x_\epsilon) \leq \epsilon_D \| c(x_\epsilon) \| \quad (46)$$

or

$$\| c(x_\epsilon) \| \leq \epsilon_P \quad \text{and} \quad \pi_\Lambda(x_\epsilon, y_\epsilon) \leq \epsilon_D \| (y_\epsilon, 1) \|, \quad (47)$$

where $\Lambda(x, y) \overset{\text{def}}{=} f(x) + y^T c(x)$ is the Lagrangian with respect to the equality constraints and y_ϵ is a vector of Lagrange multipliers associated with the equality constraints.

Proof If the ARpGC algorithm terminates in Phase 1, we immediately obtain that (46) holds, and Lemma 6 then ensures that the number of evaluations of c and its derivatives cannot exceed

$$\left\lfloor \kappa_{cc}^c \| c(x_0) \| \max \left[\epsilon_P^{-1}, \epsilon_P^{-\frac{1}{p}} \epsilon_D^{-\frac{p+1}{p}} \right] \right\rfloor + 1. \quad (48)$$

The conclusions of the theorem therefore hold in this case.

Let us now assume that termination does not occur in Phase 1. Then the ARpGC algorithm must terminate after a number of evaluations of f and c and their derivatives which is bounded above by the upper bound on the number of evaluations in Phase 1 given by (48) plus the bound on the number of evaluations of μ given by (42) and Lemma 10. Using the fact that $\lfloor a \rfloor + \lfloor b \rfloor \leq \lfloor a + b \rfloor$ for $a, b \geq 0$ and $\lfloor a + i \rfloor = \lfloor a \rfloor + i$ for $a \geq 0$ and $i \in \mathbb{N}$, this yields the combined upper bound

$$\left[\kappa_{\mathrm{CC}}^c \|c(x_0)\| \max\left[\epsilon_{\mathrm{P}}^{-1}, \epsilon_{\mathrm{P}}^{-\frac{1}{p}} \epsilon_{\mathrm{D}}^{-\frac{p+1}{p}} \right] \right.$$
$$\left. + \left[(1-\delta)\kappa_{\mathrm{CC}}^{\mu} \epsilon_{\mathrm{P}}^{\frac{p-1}{p}} \max\left[1, \epsilon_{\mathrm{P}}^{\frac{p-1}{p}} \epsilon_{\mathrm{D}}^{-\frac{p+1}{p}} \right] \right] \times \left[\frac{f(x_1) - f_{\mathrm{low}} + 1}{1-\delta} \epsilon_{\mathrm{P}}^{-1} \right] \right] + 2,$$

and (45) follows.

Remember now that (38) holds at termination of Phase 2, and therefore that

$$\epsilon_{\mathrm{P}} \geq \|r(x_\epsilon, t_\epsilon)\| \geq \delta\epsilon_{\mathrm{P}}. \tag{49}$$

Moreover, we also obtain from (38) that

$$|P_{\mathscr{G}}[x_\epsilon - J(x_\epsilon)^T c(x_\epsilon) - (f(x_\epsilon) - t_k)\nabla_x f(x_\epsilon)] - x_\epsilon\| = \pi_\mu(x_\epsilon, t_\epsilon) \leq \epsilon_{\mathrm{D}}\|r(x_\epsilon, t_\epsilon)\|. \tag{50}$$

Assume first that $f(x_\epsilon) = t_\epsilon$. Then, using the definition of $r(x, t)$, we deduce that

$$\pi_{\frac{1}{2}\|c\|^2}(x_\epsilon) = \|P_{\mathscr{G}}[x_\epsilon - J(x_\epsilon)^T c(x_\epsilon)] - x_\epsilon\| \leq \epsilon_{\mathrm{D}}\|c(x_\epsilon)\|,$$

and (46) is again satisfied because (49) gives that $\|c(x_\epsilon)\| = \|r(x_\epsilon, t_\epsilon)\| \geq \delta\epsilon_{\mathrm{P}}$.

Assume now that $f(x_\epsilon) > t_\epsilon$ (the case where $f(x_\epsilon) < t_\epsilon$ is excluded by (38)) and note that

$$0 < f(x_\epsilon) - t_\epsilon \leq \epsilon_{\mathrm{P}} \leq 1$$

because of the second bound in (36) and the decreasing nature of $\|r(x, t_k)\|$ during inner iterations. Defining now

$$y_\epsilon \stackrel{\text{def}}{=} \frac{c(x_\epsilon)}{f(x_\epsilon) - t_\epsilon},$$

and successively using Lemma 11 with $x = x_\epsilon$, $v = -(J(x_\epsilon)^T c(x_\epsilon) + (f(x_\epsilon) - t_\epsilon)\nabla_x f(x_\epsilon))$ and $\tau = 1/(f(x_\epsilon) - t_\epsilon) \geq 1$, the third part of (25), (49) and the definition of $r(x, t)$, we deduce that

$$\begin{aligned}
\pi_\Lambda(x_\epsilon, y_\epsilon) &= \|P_{\mathscr{G}}[x_\epsilon - J(x_\epsilon)^T \frac{c(x_\epsilon)}{f(x_\epsilon) - t_\epsilon} - \nabla_x f(x_\epsilon)] - x_\epsilon\| \\
&\leq \frac{1}{f(x_\epsilon) - t_\epsilon} \|P_{\mathscr{G}}[x_\epsilon - J(x_\epsilon)^T c(x_\epsilon) - (f(x_\epsilon) - t_\epsilon)\nabla_x f(x_\epsilon)] - x_\epsilon\| \\
&= \frac{1}{f(x_\epsilon) - t_\epsilon} \pi_\mu(x_\epsilon, t_\epsilon) \\
&\leq \epsilon_{\mathrm{D}} \frac{\|r(x_\epsilon, t_\epsilon)\|}{f(x_\epsilon) - t_\epsilon} \\
&= \epsilon_{\mathrm{D}}\|(y_\epsilon, 1)\|.
\end{aligned}$$

This implies (47) since $\|c(x_\epsilon)\| \leq \|r(x_\epsilon, t_\epsilon)\| \leq \epsilon$. $\diamond$

Note that the bound (45) is $O(\epsilon^{-\frac{p+2}{p}})$ whenever $\epsilon_P = \epsilon_D = \epsilon$.

It is important to note that the complexity bound given by Theorem 2 depends on $f(x_1)$, the value of the objective function at the end of Phase 1. Giving an upper bound on this quantity is in general impossible, but can be done in some case. A trivial bound can of course be obtained if $f(x)$ is bounded above in $\mathscr{C}_\beta$. This has the advantage of providing a complexity result which is self-contained (in that it only involves problem-dependent quantities), but it is quite restrictive as it excludes, for instance, problems only involving equality constraints ($\mathscr{F} = \mathbb{R}^n$) and coercive objective functions. A bound is also readily obtained if the set $\mathscr{F}$ is itself bounded (for instance, when the variables are subject to finite lower and upper bounds) or if one assumes that the iterates generated by Phase 1 remain bounded. This may, for example, be the case if the set $\{x \in \mathbb{R}^n \mid c(x) = 0\}$ is bounded. An ϵ_P-dependent bound can finally be obtained if one is ready to assume that all derivatives of order 1 to p of $c(x)$ (and thus of $v(x)$) are bounded by a constant in the level set $\mathscr{C}_0 \stackrel{\text{def}}{=} \{x \in \mathscr{F} \mid \|c(x)\| \leq \|c(x_0)\|\}$ because it can then be shown that $\|s_k\|$ is uniformly bounded above and hence that $\|x_1 - x_0\|$ is itself bounded above by a constant times the (ϵ_P-dependent) number of iterations in Phase 1 given by Lemma 6. Using the boundedness of the gradient of $v(x)$ on the path of iterates then ensures the (extremely pessimistic) upper bound $f(x_1) = f(x_0) + O\left(\max\left[\epsilon_P^{-1}, \epsilon_P^{-\frac{1}{p}}\epsilon_D^{-\frac{p+1}{p+1-q}}\right]\right)$. Substituting this bound in (45) in effect squares the complexity of obtaining (x_ϵ, t_ϵ).

4 Discussion

We have first shown in Section 2 that, if derivatives of the objective function up to order p can be evaluated and if the p-th one is Lipschitz continuous, then the ARpCC algorithm applied to the convexly constrained problem (1) needs at most $O(\epsilon^{-\frac{p+1}{p}})$ evaluations of f and its derivatives to compute an ϵ-approximate first-order critical point. This worst-case bound corresponds to that obtained in [4] when $p = 2$, but with significantly weaker assumptions. Indeed, the present proposal no longer needs any assumption on the number of descent steps in the subproblem solution, the iterates are no longer assumed to remain in a bounded set and the Lipschitz continuity of the gradient is no longer necessary. That these stronger results are obtained as the result of a considerably simpler analysis is an added bonus. While we have not developed here the case (covered for $p = 2$ in [4]), where the p-th derivative is only known approximately (in the sense that $\nabla_x^p f(x_k)$ is replaced in the model's expression by some tensor B_k such that the norm of $(\nabla_x^p f(x_k) - B_k)$ applied $p - 1$ times to s_k must be $O(\|s_k\|^p)$), the generalization of the present proposal to cover this situation is easy.

The proposed worst-case evaluation bound also generalizes that of [3] for unconstrained optimization to the case of set-constrained problems, under very weak assumptions on the feasible set. As was already the case for $p \leq 2$, it is remarkable that the complexity bound for the considered class of problems (which includes the standard bound constrained case) is, for all $p \geq 1$, identical in order to that of unconstrained problems.

The present framework for handling convex constraints is however not free of limitations, resulting from the choice to transfer difficulties associated with the original problem to the subproblem solution, thereby sparing precious evaluations of f and its derivatives. The first is that we need to compute projections onto the feasible set to obtain values of π_f and π_{m_k}. While this is straightforward and computationally inexpensive for simple convex sets such boxes, spheres, cylinders or the order-simplex, the process might be more intensive for the general case. The second limitation is that, even if the projections can be computed, the approximate solution of the subproblem may also be very expensive in terms of internal calculations (we do not consider here suitable algorithms for this purpose). Observe nevertheless that, crucially, neither the computation of the projections nor the subproblem solution involves evaluating the objective function or its derivatives: despite their potential computational drawbacks, they have therefore no impact on the evaluation complexity of the original problem. However, as the cost of evaluating any constraint function/derivative possibly necessary for computing projections is neglected by the present approach, it must therefore be seen as a suitable framework to handle "cheap inequality constraint" such as simple bounds.

We have also shown in Section 3 that the evaluation complexity of finding an approximate first-order scaled critical point for the general smooth nonlinear optimization problem involving both equality and inequality constraints is at most $O(\epsilon_P^{-1/p} \epsilon_D^{-(p+1)/p})$ evaluations of the objective function, constraints and their derivatives up to order p. We refer here to an "approximate scaled critical point" in that such a point is required to satisfy (46) or (47), where the accuracy is scaled by the size of the constraint violation or that of the Lagrange multipliers. Because this bound now only depends on the assumptions necessary to prove the evaluation complexity bound for the ARpCC algorithm in Section 2, it therefore strengthens and generalizes that of [7] since the latter directly hinges on [4]. Moreover, it also corrects an unfortunate error[1] in [7] that allows a vector of Lagrange multipliers whose sign is arbitrary (in line with a purely first-order setting where minimization and maximization are not distinguished). The present analysis now yields the multiplier with the sign associated with minimization.

Interestingly, an $O(\epsilon_P \epsilon_D^{-(p+1)/p} \min[\epsilon_D, \epsilon_P]^{-(p+1)/p})$ evaluation complexity bound was also proved by Birgin et al. in [2] for *unscaled*, standard KKT conditions and in the least expensive of three cases depending on the degree of degeneracy

---

[1] The second equality in the first equation of Lemma 3.4 in [7] only holds if one is ready to flip the gradient's sign if necessary.

identifiable by the algorithm.[2] Even if the bounds for the scaled and unscaled cases coincide in order when $\epsilon_P \leq \epsilon_D$, comparing the two results is however not straightforward. On one hand the scaled conditions take into account the possibly different scaling of the objective function and constraints. On the other hand the same scaled conditions may result in earlier termination with (47) if the Lagrange multipliers are very large, as (47) is then consistent with the weaker requirement of finding a John's point. But the framework discussed in the present paper also differs from that of [2] in two additional significant ways. The first is that the present one provides a potentially stronger version of the termination of the algorithm at infeasible points (in Phase 1): indeed the second part of (46) can be interpreted as requiring that the size of the feasible gradient of $\|c(x)\|$ is below ϵ_D, while [2] considers the gradient of $\|c(x)\|^2$ instead. The second is that, if termination occurs in Phase 2 for an x_ϵ such that $\pi_{\frac{1}{2}\|c\|^2}(x_\epsilon) = \|J(x_\epsilon)^T c(x_\epsilon)\|$ is itself of order $\epsilon_P \epsilon_D$ (thereby covering the case, where $f(x_\epsilon) = t_k$ discussed in Theorem 2), then Birgin et al. show that the Łojasiewicz inequality [10] must fail for c in the limit for ϵ_P and ϵ_D tending to zero (see [2] for details). This observation is interesting because smooth functions satisfy the Łojasiewicz inequality under relatively weak conditions, implying that termination in these circumstances is unlikely. The same information is also obtained in [2], albeit at the price of worsening the evaluation complexity bound mentioned above by an order of magnitude in ϵ_D. We also note that the approach of [2] requires the minimization, at each iteration, of a residual whose second derivatives are discontinuous, while all functions used in the present paper are p times continuously differentiable. A final difference between the two approaches is obviously our introduction of π_A and $\pi_{\frac{1}{2}\|c\|^2}$ in the expression of the criticality condition in Theorem 2 for taking the inequality constraints into account.

We conclude our discussion by a remark about criticality measures. At variance with [4] and [7], we have preferred, in this paper, to use the first-order criticality measure $\pi_f(x)$ rather than $\chi_f(x)$, the decrease in the linearized function in the intersection of the feasible set and the ball of radius one. While a similar result can indeed be obtained for this latter measure (in this case for general closed non-empty convex sets even in Section 3), our choice is motivated by the observation made by Gratton, Mouffe and Toint [9] that π_f is backward stable as a criticality measure (in the sense that an approximate solution of problem (1) yielding a small value of π_f can be interpreted as the exact solution of a neighbouring problem), while this is not the case for χ_f even when $\mathscr{F}$ is described by simple bounds on the problem's variables.

Acknowledgements The work of the second author was supported by EPSRC grants EP/I013067/1 and EP/M025179/1. The third author gratefully acknowledges the financial support of the Belgian Fund for Scientific Research, the Leverhulme Trust and Balliol College (Oxford).

---

[2]This result also assumes boundedness of $f(x_1)$.

References

1. Bertsekas, D.P.: Nonlinear Programming, 4th edn. Athena Scientific, Belmont (1999)
2. Birgin, E.G., Gardenghi, J.L., Martínez, J.M., Santos, S.A., Toint, Ph.L.: Evaluation complexity for nonlinear constrained optimization using unscaled KKT conditions and high-order models. SIAM J. Optim. **26**(2), 951–967 (2016)
3. Birgin, E.G., Gardenghi, J.L., Martínez, J.M., Santos, S.A., Toint, Ph.L.: Worst-case evaluation complexity for unconstrained nonlinear optimization using high-order regularized models. Math. Program. **163**(1–2), 359–368 (2017)
4. Cartis, C., Gould, N.I.M., Toint, Ph.L.: An adaptive cubic regularization algorithm for nonconvex optimization with convex constraints and its function-evaluation complexity. IMA J. Numer. Anal. **32**(4), 1662–1695 (2012)
5. Cartis, C., Gould, N.I.M., Toint, Ph.L.: On the complexity of finding first-order critical points in constrained nonlinear optimization. Math. Program. A **144**(1), 93–106 (2013)
6. Cartis, C., Gould, N.I.M., Toint, Ph.L.: On the evaluation complexity of cubic regularization methods for potentially rank-deficient nonlinear least-squares problems and its relevance to constrained nonlinear optimization. SIAM J. Optim. **23**(3), 1553–1574 (2013)
7. Cartis, C., Gould, N.I.M., Toint, Ph.L.: On the evaluation complexity of constrained nonlinear least-squares and general constrained nonlinear optimization using second-order methods. SIAM J. Numer. Anal. **53**(2), 836–851 (2015)
8. Cartis, C., Gould, N.I.M., Toint, Ph.L.: Improved worst-case evaluation complexity for potentially rank-deficient nonlinear least-Euclidean-norm problems using higher-order regularized models. Technical Report NA-TR 15–17, Numerical Analysis Group, Mathematical Institute, University of Oxford (2015)
9. Gratton, S., Mouffe, M., Toint, Ph.L.: Stopping rules and backward error analysis for bound-constrained optimization. Numer. Math. **119**(1), 163–187 (2011)
10. Łojasiewicz, S.: Ensembles semi-analytiques. Technical report, Institut des Hautes Etudes Scientifiques, Bures-sur-Yvette, France (1965). Available online at http://perso.univ-rennes1.fr/michel.coste/Lojasiewicz.pdf
11. Nesterov, Yu.: Introductory Lectures on Convex Optimization. Applied Optimization. Kluwer Academic Publishers, Dordrecht (2004)

Data-Dependent Approximation in Social Computing

Weili Wu, Yi Li, Panos M. Pardalos ⓘ, and Ding-Zhu Du

Abstract Data-dependent approximation is a new approach for the study of nonsubmodular optimization problems. This approach has attracted a lot of research especially in the area of social computing, where nonsubmodular combinatorial optimization problems have been recently formulated. In this chapter, we present some theoretical results on the data-dependent approximation approach. In addition, some related open problems are discussed.

1 Introduction

Online social networks (including sites like FaceBook, LinkedIn, ResearchGate, and messengers like Skype) are among the most popular sites and communication tools on the Internet. The users of these sites and tools form huge social networks. Their activities provide a huge platform for research of social computing, especially the study on social influence [1, 2, 7, 9, 13, 23, 25–27] which have many applications, including in marketing [11, 17] and presidential election, which received a lot of attentions.

There are many combinatorial optimization problems raised from those activities. A lot of them have nonlinear objective functions with discrete structure, which give a lot of motivations for studying monotone submodular optimization [15, 18], nonmonotone submodular optimization [5, 6, 10], and nonsubmodular optimization [8, 14, 24]. In particular, the nonsubmodular optimization is a recent hot research direction since no algorithm has been found to produce a solution with a theoretical guaranteed performance in traditional standard [4].

W. Wu · Y. Li · D.-Z. Du (✉)
Department of Computer Science, University of Texas at Dallas, Richardson, TX, USA
e-mail: weiliwu@utdallas.edu; yi.li@utdallas.edu; dzdu@utdallas.edu

P. M. Pardalos
Department of Industrial & Systems Engineering, University of Florida, Gainesville, FL, USA
e-mail: pardalos@ufl.edu

© Springer Nature Switzerland AG 2019
I. C. Demetriou, P. M. Pardalos (eds.), *Approximation and Optimization*,
Springer Optimization and Its Applications 145,
https://doi.org/10.1007/978-3-030-12767-1_3

How to give a solution with reasonable analysis? Recently, Lu et al. [12] proposed a new approach, called *data-dependent* approximation which attracts a lot of attentions. In this article, we try to give some theoretical foundation for this approach and identify some theoretical open problems.

To start our study, let us first recall some basic concepts and terminologies. Consider a set function $f : 2^X \to R$. f is called a *submodular* function if for any two sets $A \subset B$ and any element $x \in X \setminus B$,

$$\Delta_x f(A) \geq \Delta_x f(B)$$

where $\Delta_x f(A) = f(A \cup \{x\}) - f(A)$. f is called a *monotone nondecreasing* function if for any two sets $A \subset B$, $f(A) \leq f(B)$, i.e., for any element $x \in B \setminus A$,

$$\Delta_x f(A) \geq \Delta_x f(B).$$

2 Example

Let us first introduce the data-dependent approximation through an example.

Example (Activity Maximization [3]) Consider a social network represented by a directed graph $G = (V, E)$, together with an information diffusion model m. In this model, each node has two states, active and inactive. Initially, all nodes are in inactive state. The influence diffusion consists of discrete steps. At the beginning, a set of nodes are activated. Nodes in this set are called *seeds*. At each subsequent step, every inactive node v evaluates its status and decides whether it should be activated or not, based on the rule in the model m. The process ends at a step in which no more inactive node becomes active.

Let S denote the set of seeds and $I_m(S)$ the set of active nodes at the end of diffusion process. Suppose that for each par of active nodes $u, v \in I_m(S)$, if (u, v) is an edge of G, i.e., $(u, v) \in E$, then an activity profit $A(u, v)$ will be generated where $A : E \to R_+$ is a nonnegative activity profit function. The activity maximization is the following problem:

$$(\alpha) \qquad \max \ \alpha(S) = \sum_{(u,v)\in E: u, v \in I_m(S)} A(u, v)$$

$$\text{subject to } |S| \leq k,$$

$$S \subseteq V.$$

This problem has been proved to be NP-hard in [3]. There are also counterexamples in [3], which show that $\alpha(S)$ is neither submodular nor supermodular. However, Lu et al. [3] introduced two monotone nondecreasing submodular set functions $\beta : 2^V \to R_+$ and $\gamma : 2^V \to R_+$ such that for any $S \in 2^V$, $\beta(S) \geq \alpha(S) \geq \gamma(S)$.

These two set functions are defined as follows:

$$\beta(S) = \sum_{(u,v)\in E:u\in I_m(S)} A(u,v)$$

and

$$\gamma(S) = \sum_{s\in S} \sum_{(u,v)\in E:u,v\in I_m(\{s\})} A(u,v).$$

By a theorem of Nemhauser and Wolsey [15], there is a greedy algorithm which is able to find $(1 - e^{-1})$-approximation solutions for the following two problems.

$$(\beta) \qquad \max \ \beta(S)$$
$$\text{subject to} \ |S| \leq k,$$
$$S \subseteq V;$$

$$(\gamma) \qquad \max \ \gamma(S)$$
$$\text{subject to} \ |S| \leq k,$$
$$S \subseteq V.$$

Let S_β and S_γ be $(1 - e^{-1})$-approximation solutions for problem (β) and (γ), respectively. Let S_α be a feasible solution for problem (α). Choosing the best one from S_α, S_β, and S_γ, we would obtain a data-dependent approximation solution for problem (α), i.e., the data-dependent approximation solution is

$$S_{data} = \text{argmax}_{S\in\{S_\alpha,S_\beta,S_\gamma\}} \alpha(S).$$

For this solution, we have the following theoretical guaranteed performance.

Theorem 1

$$\alpha(S_{data}) \geq (1 - e^{-1}) \cdot \max\left(\frac{\alpha(S_\beta)}{\beta(S_\beta)}, \frac{opt_\gamma}{opt_\alpha}\right) \cdot opt_\alpha$$

where opt_α (opt_γ) is the objective function value of an optimal solution for problem (α) (problem (γ)).

Proof First, we have

$$\alpha(S_\beta) = \frac{\alpha(S_\beta)}{\beta(S_\beta)} \cdot \beta(S_\beta)$$
$$\geq \frac{\alpha(S_\beta)}{\beta(S_\beta)} \cdot (1 - e^{-1}) \cdot opt_\beta$$

$$\geq \frac{\alpha(S_\beta)}{\beta(S_\beta)} \cdot (1 - e^{-1}) \cdot \beta(OPT_\beta)$$

$$\geq \frac{\alpha(S_\beta)}{\beta(S_\beta)} \cdot (1 - e^{-1}) \cdot opt_\alpha,$$

where OPT_α is an optimal solution for problem (α) and opt_β is the objective function value of an optimal solution for problem (β). Secondly, we have

$$\alpha(S_\gamma) \geq \gamma(S_\gamma)$$
$$\geq (1 - e^{-1}) \cdot opt_\gamma$$
$$= (1 - e^{-1}) \cdot \frac{opt_\gamma}{opt_\alpha} \cdot opt_\alpha.$$

Therefore, the theorem holds.   □

In general, suppose we study an optimization problem $\max\{\alpha(x) \mid x \in \Omega\}$ where $\alpha(\cdot)$ is a set function. Then we may find a submodular upper bound $\beta(x)$ and a submodular lower bound $\gamma(x)$ for $\alpha(x)$. Suppose we can find an η-approximation solution for $\max\{\beta(x) \mid x \in \Omega\}$ and a τ-approximation solution for $\max\{\gamma(x) \mid x \in \Omega\}$. Then we may find a data-dependent approximation solution in the following way.

Data-Dependent Approximation
Compute η-approximation x_β for $\max\{\beta(x) \mid x \in \Omega\}$.
Compute τ-approximation x_γ for $\max\{\gamma(x) \mid x \in \Omega\}$.
Compute a feasible solution x_α for $\max\{\alpha(x) \mid x \in \Omega\}$.
Choose $x_{data} = \text{argmax}(\alpha(x_\alpha), \alpha(x_\beta), \alpha(x_\gamma))$.

This solution x_{data} has the following performance.

Theorem 2

$$\alpha(x_{data}) \geq \max\left(\eta \cdot \frac{\alpha(x_\beta)}{\beta(x_\beta)}, \tau \cdot \frac{opt_\gamma}{opt_\alpha}\right) \cdot opt_\alpha$$

where opt_α (opt_γ) is the objective function value of an optimal solution for problem $\max\{\alpha(x) \mid x \in \Omega\}$ (problem $\max\{\gamma(x) \mid x \in \Omega\}$).

To end this section, let us make two remarks. First, note that the feasible solution S_α or x_α does not play any role in the establishment of the approximation performance of the data-approximation. Why do we need it? Actually, S_α may improve the performance in computer simulation. Since it can be arbitrarily selected, we may choose any selection method. For example, in the activity maximization, the set function $\alpha(S)$ is monotone nondecreasing, we may use the following greedy algorithm to select S_α.

Greedy Algorithm
$S_0 \leftarrow \emptyset$;
for $i = 1$ **to** k **do**
$\quad x = \text{argmax}_{x \in V \setminus S_{i-1}} (\alpha(S_{i-1} \cup \{x\}) - \alpha(S_{i-1}))$ and
$\quad S_i \leftarrow S_{i-1} \cup \{x\}$;
output $S_\alpha = S_k$.

The second remark is about the objective function $\alpha(\cdot)$. In many problems raised from social computing, since the information diffusion model is probabilistic, such as the linear threshold model and the independent cascade model, the function $\alpha(\cdot)$ is the expectation of profit. For example, in the activity maximization, $\alpha(\cdot)$ would become

$$\alpha(S) = E[\sum_{(u,v) \in E: u, v \in I_m(S)} A(u, v)].$$

Correspondingly, the upper bound $\beta(S)$ and the lower bound $\gamma(S)$ have to be modified into expectations as follows:

$$\beta(S) = E[\sum_{(u,v) \in E: u \in I_m(S)} A(u, v)]$$

and

$$\gamma(S) = E[\sum_{s \in S} \sum_{(u,v) \in E: u, v \in I_m(\{s\})} A(u, v)].$$

In this case, it may be hard to compute $\beta(S)$ and $\gamma(S)$ for a given S. For example, with the linear threshold model or the independent cascade model, computing $\beta(S)$ and $\gamma(S)$ for given S are #P-hard problems. This would make the greedy algorithm in [15] unable to be implemented in polynomial-time. Instead, we have to employ randomized approximation algorithm in [16, 19, 20, 22]. Since $\beta(S)$ and $\gamma(S)$ are still monotone nondecreasing and submodular, for any constants $\varepsilon > 0$ and $\rho > 0$, there exists a polynomial-time randomized algorithm which with probability at least $1 - \rho$ produces a solution with approximation performance ratio $1 - e^{-1} - \varepsilon$ for problem (β) (or problem (γ)). Then the approximation performance of the data-dependent approximation has to be modified correspondingly.

3 Theoretical Notes

For any set function $\alpha : 2^X \rightarrow R$, is it always able to find a pair of monotone nondecreasing submodular functions $\beta : 2^X \rightarrow R$ and $\gamma : 2^X \rightarrow R$ which are an upper bound and a lower bound for $\alpha(\cdot)$, respectively? The answer for this theoretical question is yes.

Following result can be found in [8].

Lemma 1 *Any set function $\alpha : 2^X \to R$ can be expressed as a difference of two monotone nondecreasing submodular functions, i.e., there exists a pair of monotone nondecreasing submodular set functions $\beta : 2^X \to R$ and $\zeta : 2^X \to R$ such that $\alpha(S) = \beta(S) - \zeta(S), \forall S \in 2^X.$*

With this result, we can prove the following.

Theorem 3 *For any set function $\alpha : 2^X \to R$, there exists a pair of monotone nondecreasing submodular functions $\beta : 2^X \to R$ and $\gamma : 2^X \to R$ such that $\beta(S) \geq \alpha(S) \geq \gamma(S) \; \forall S \subseteq X.$*

Proof First, we remark that the functions $\beta(\cdot)$ and $\zeta(\cdot)$ in Lemma 1 can further be required to be nonnegative. In fact, since they are monotone nondecreasing, we have

$$\beta(\emptyset) = \min_{S \subseteq X} \beta(S) \text{ and } \zeta(\emptyset) = \min_{S \subseteq X} \zeta(S).$$

Set $c = \max(|\beta(\emptyset)|, |\gamma(\emptyset)|)$ and define $\beta'(S) = \beta(S) + c$ and $\zeta'(S) = \zeta(S) + c$ for any $S \subseteq X$. Then c is a constant and moreover, $\beta'(\cdot)$ and $\zeta'(\cdot)$ are nonnegative, monotone nondecreasing, and submodular functions such that for any $S \subseteq X$, $\alpha(S) = \beta'(S) - \zeta'(S)$.

Clearly, $\beta'(\cdot)$ is an upper bound for function $\alpha(\cdot)$. To obtain a lower bound, we set $\gamma(S) = \beta'(S) - \zeta'(X)$. It is easy to see that $\alpha(S) \geq \gamma(S)$ for any $S \subseteq X$. Since $\zeta'(X)$ is a constant, $\gamma(\cdot)$ is monotone nondecreasing and submodular. $\square$

For this theorem, we may note the following:

1. The proof of Lemma 1 in [8] is not constructive. Therefore, our proof of Theorem 3 is not constructive, neither.
2. There are infinitely many pair of an upper bound and a lower bound meeting requirement in Theorem 3. In fact, simply choose any constant $c' > c$ and replace c by c'. We would get a new pair.

From fact 2, we may feel that it is necessary to set up a measure and the quality of such a pair. A naive measure is

$$Q(\beta, \gamma) = \max_{S \subseteq X}(\beta(S) - \gamma(S)).$$

Motivated from this measure and fact 1, we propose the following open problems.

Open Problem 2 *For any set function $\alpha : 2^X \to R$ and a constant $q > 0$, is there a pair of monotone nondecreasing submodular functions $\beta : 2^X \to R$ and $\gamma : 2^X \to R$ such that $\beta(S) \geq \alpha(S) \geq \gamma(S)$ for any $S \subseteq X$ and $Q(\beta, \gamma) \leq q$?*

Open Problem 3 *If Open Problem 2 receives No-answer, then is there an efficient method to tell whether such a pair of upper and lower bounds exists or not?*

4 Conclusion

Our theoretical notes indicate that the data-dependent approximation exists for any nonsubmodular maximization problem. A similar study can show the existence of data-dependent approximation for any nonsubmodular minimization problem. However, to improve the quality of the data-dependent approximation, we may need a lot of efforts on the DS function maximization where a DS function is a difference of two submodular functions. Some fundamental theoretical problems are still open. For more information on social networks, please refer to [21].

References

1. Bharathi, S., Kempe, D., Salek, M.: Competitive influence maximization in social networks. In: Deng, X., Graham, F.C. (eds.) Internet and Network Economics. WINE 2007. Lecture Notes in Computer Science, vol. 4858, pp. 306–311. Springer, Berlin (2007)
2. Chen, W., Wang, C., Wang, Y.: Scalable influence maximization for prevalent viral marketing in large-scale social networks. In: ACM SIGKDD International Conference on Knowledge Discovery and Data Mining, pp. 1029–1038 (2010)
3. Chen, W., Lin, T., Tan, Z., Zhao, M., Zhou, X.: Robust influence maximization. In: ACM SIGKDD International Conference on Knowledge Discovery and Data Mining, pp. 795–804 (2016)
4. Du, D., Ko, K., Hu, X.: Design and Analysis of Approximation Algorithms. Springer, New York (2012)
5. Feige, U., Mirrokni, V.S., Vondrák, J.: Maximizing non-monotone submodular functions. SIAM J. Discret. Math. **23**(4), 2053–2078 (2007)
6. Feldman, M., Naor, J., Schwartz, R.: A unified continuous greedy algorithm for submodular maximization. In: IEEE Symposium on Foundations of Computer Science, pp. 570–579 (2011)
7. Goyal, A., Bonchi, F., Lakshmanan, L.V.S.: A data-based approach to social influence maximization. Proc. VLDB Endow. **5**(1), 2011 (2012)
8. Iyer, R., Bilmes, J.: Algorithms for approximate minimization of the difference between submodular functions, with applications. In Twenty-Eighth Conference on Uncertainty in Artificial Intelligence, pp. 407–417 (2012)
9. Kempe, D., Kleinberg, J., Tardos, É.: Maximizing the spread of influence through a social network. Theory of Computing **11**, 137–146 (2010)
10. Lee, J., Mirrokni, V.S., Nagarajan, V., Sviridenko, M.: Non-monotone submodular maximization under matroid and knapsack constraints. In: ACM Symposium on Theory of Computing, pp. 323–332 (2009)
11. Leskovec, J., Adamic, L.A., Huberman, B.A.: The dynamics of viral marketing. ACM Trans. Web **1**(1), 39 (2007)
12. Lu, W., Chen, W., Lakshmanan, L.V.S.: From competition to complementarity: comparative influence diffusion and maximization. Proc. VLDB Endow. **9**(2), 60–71 (2015)
13. Mossel, E., Roch, S.: On the submodularity of influence in social networks. In: Thirty-Ninth ACM Symposium, pp. 128–134 (2007)
14. Narasimhan, M., Bilmes, J.A.: A submodular-supermodular procedure with applications to discriminative structure learning. In: Proceedings of the Twenty-First Conference on Uncertainty in Artificial Intelligence, pp. 404–412 (2012)
15. Nemhauser, G.L., Wolsey, L.A.: Best algorithms for approximating the maximum of a submodular set function. Math. Oper. Res. **3**(3), 177–188 (1978)

16. Nguyen, H.T., Thai, M.T., Dinh, T.N.: Stop-and-stare: optimal sampling algorithms for viral marketing in billion-scale networks. In: International Conference on Management of Data, pp. 695–710 (2016)
17. Richardson, M., Domingos, P.: Mining knowledge-sharing sites for viral marketing. In: Eighth ACM SIGKDD International Conference on Knowledge Discovery and Data Mining, pp. 61–70 (2002)
18. Sviridenko, M.: A note on maximizing a submodular set function subject to a knapsack constraint. Oper. Res. Lett. **32**(1), 41–43 (2004)
19. Tang, Y., Xiao, X., Shi, Y.: Influence maximization: near-optimal time complexity meets practical efficiency. In: ACM Sigmod International Conference on Management of Data, pp. 75–86 (2014)
20. Tang, Y., Shi, Y., Xiao, X.: Influence maximization in near-linear time: a martingale approach. In: ACM SIGMOD International Conference on Management of Data, pp. 1539–1554 (2015)
21. Thai, M.T., Pardalos, P.M.: Handbook of Optimization in Complex Networks. Springer, New York (2012)
22. Tong, G., Wu, W., Guo, L., Li, D., Liu, C., Liu, B., Du, D.: An efficient randomized algorithm for rumor blocking in online social networks. IEEE Trans. Netw. Sci. Eng. **PP**(99), 1–1 (2017)
23. Wang, A., Wu, W., Cui, L.: On Bharathi–Kempe–Salek conjecture for influence maximization on arborescence. J. Comb. Optim. **31**(4), 1678–1684 (2016)
24. Wu, C., Wang, Y., Lu, Z., Pardalos, P.M., Xu, D., Zhang, Z., Du, D.: Solving the degree-concentrated fault-tolerant spanning subgraph problem by DC programming. Math. Program. **169**(1), 1–21 (2018)
25. Xu, W., Lu, Z., Wu, W., Chen, Z.: A novel approach to online social influence maximization. Soc. Netw. Anal. Min. **4**(1), 1–13 (2014)
26. Zhang, H., Dinh, T.N., Thai, M.T.: Maximizing the spread of positive influence in online social networks. In: IEEE International Conference on Distributed Computing Systems, pp. 317–326 (2013)
27. Zhang, Z., Lu, Z., Wu, W.: Solution of Bharathi-Kempe-Salek conjecture on influence maximization in arborescence. J. Comb. Optim. **33**, 803–808 (2016)

Multi-Objective Evolutionary Optimization Algorithms for Machine Learning: A Recent Survey

Stamatios-Aggelos N. Alexandropoulos, Christos K. Aridas, Sotiris B. Kotsiantis, and Michael N. Vrahatis

Abstract The machine learning algorithms exploit a given dataset in order to build an efficient predictive or descriptive model. Multi-objective evolutionary optimization assists machine learning algorithms to optimize their hyper-parameters, usually under conflicting performance objectives and selects the best model for a given task. In this paper, recent multi-objective evolutionary approaches for four major data mining and machine learning tasks, namely: (a) data preprocessing, (b) classification, (c) clustering, and (d) association rules, are surveyed.

1 Introduction

For a given optimization task, in general, optimization consists of the following main issues [35]:

(a) *Objective function*: the quantity to be optimized (maximized or minimized).
(b) *Variables*: the inputs to the objective function.
(c) *Constraints*: the restrictions assigned to the inputs of the objective function.

Therefore, the purpose of an optimizer is to determine properly the values to the inputs of the objective function, in such a way to attain the optimal solution for the given function and fulfilling all the required constraints.

Various real-world optimization tasks often suffer from the following difficulties [112]:

1. In many cases, it is difficult to discern the global optimal minimizers from the local optimal ones.

S.-A. N. Alexandropoulos · C. K. Aridas · S. B. Kotsiantis · M. N. Vrahatis (✉)
Computational Intelligence Laboratory – CILab, Department of Mathematics, University of
Patras, Patras, Greece
e-mail: alekst@math.upatras.gr; char@upatras.gr; sotos@math.upatras.gr;
vrahatis@math.upatras.gr

© Springer Nature Switzerland AG 2019
I. C. Demetriou, P. M. Pardalos (eds.), *Approximation and Optimization*,
Springer Optimization and Its Applications 145,
https://doi.org/10.1007/978-3-030-12767-1_4

2. The evaluation of the solutions may be difficult in the presence of noise.
3. The search space may be large, so the dimensionality of the problem grows similarly. This causes the so-called curse of dimensionality problem.
4. Difficulties associated with the given limitations assigned to the inputs of the objective function.
5. Necessity for problem-specific optimization techniques.

In the case where the quantity to be optimized is expressed by only one objective function, the problem is referred to as a *uni-objective* or *single-objective* problem. While, a *multi-objective* problem identifies more than one individual targets (sub-objectives) that should be optimized at the same time.

Various applications [44, 62, 85, 119] of machine learning and other types of problems [55, 63, 91] have been handled by techniques [6, 47, 108, 113] that belong to the field of machine learning, require the fulfillment of various conditions. The simultaneous fulfillment of such conditions, as well as the optimization of the parameters incorporated in machine learning methods, constitutes a difficult optimization problem. Indeed, the majority of these algorithms [20, 21, 132] require the optimization of multiple objectives, so that the outcome result to be reliable and competitive. For instance, in feature selection task, the desired feature set has to be the minimum set that maximizes the performance of the classifier. Therefore, two conditions are required:

(a) A minimum subset of features, and
(b) These features should maximize the performance of the algorithm.

Hence, the majority of learning problems are multi-objective in nature and thus, it is evident to consider learning problems as multi-objective ones. Freitas [37] presented the above-described purpose, that should be simultaneously optimized by certain conditions, so that the performance of the building model to be eventually high. As for the most part, the optimization of a number of parameters is required, in order for the accuracy of the model to be maximized. To achieve this goal, there are three different approaches, namely:

1. The conversion of the initial *multi-objective* problem into *single-objective* one by using properly a *weighted approach*.
2. The *lexicographical approach*, where the objectives are prioritized.
3. The well-known and widely used *Pareto approach*, which gives a whole set of non-dominated solutions.

An important issue provided by a multi-objective optimization algorithm [92, 118, 130] is that, instead of one solution, to return a set of good "candidate" solutions, the so-called *non-dominated solutions*, which the user can compare with one another in order to select the most appropriate one for the given purpose [22]. The set of solutions that the algorithm returns represents the best possible trade-offs among the objectives. However, the *Pareto principle* [90] indicates that the relationship between inputs and outputs is not balanced. Thus, the *Pareto rule* (or 80/20 rule) [82] "obeys" to a distribution, known as *Pareto distribution*, which reveals that 20% of the invested inputs are responsible for 80% of the obtained results.

In many cases, the decision of an expert, the so-called *decision maker* [56], plays a key role. Thus, the opinion of the decision maker may be requested initially before the beginning of the solution process. According to the information provided by the expert, the most appropriate solution is required to meet the conditions that have been set. However, the opinion of an expert can be requested after finding a number of appropriate solutions. In this case, the expert selects the best among them. In addition, the expert's opinion may take place during the process. Specifically, the model iteratively asks for the expert's opinion in order to improve the solutions and eventually returns the required optimal set of solutions.

Despite the fact that most of the real-world problems require the optimization of multiple objectives [15], there is always an effort to reduce the number of objectives to a minimum [17]. This occurs due to the fact that the more objectives are required to be optimized, the more solutions will be. Consequently, the dimensionality and the complexity of the problem are increased and therefore the problem becomes more difficult to be solved. For an analysis of the different types of multi-objective techniques, the reader can reach more details in [58].

In this paper, the references are focused on recent published refereed journals, books, and conference proceedings. In addition, various references regarding the original work that has emanated for tackling the particular line of research under discussion are incorporated. According to this purpose, the first major attempts that have surveyed a various multi-objective evolutionary approach are appeared in [77, 78].

As it is expected, it is not possible for a single work to cover extensively all the aspects of the *multi-objective evolutionary optimization* algorithms. However, we hope through this recent review work and the most recent references that the reader will be informed on the latest interests of the scientific community on these subjects.

The following section provides the necessary background material and basic concepts of multi-objective optimization. Section 3 covers a very important aspect of machine learning, which is the data preprocessing. To this end, various multi-objective evolutionary algorithms that tackled the most common and widely used data cleaning steps are presented. The classification task and the basic models used to handle this in accordance with multi-objective optimization algorithms are described in Sect. 4. Next, in Sect. 5, cluster analysis and association rules are presented. In Sect. 6 a few of the most recent applications regarding the multi-objective evolutionary optimization algorithms are given. The paper ends in Sect. 7 with a synopsis and a short discussion.

2   Basic Concepts of Multi-Objective Optimization

The *multi-objective optimization* (MO) also named *multiple criteria optimization* handles problems where different objectives must be optimized simultaneously. For this kind of problems, *Pareto optimality* replaces the optimality notion of single-objective optimization and each Pareto optimal solution represents a trade-off of the objective functions. Hence, two solutions may obtain the same fitness value and it is desirable to obtain the largest possible count of solutions with different inherent properties.

Suppose that $\mathscr{S} \subset \mathbb{R}^n$ is an n-dimensional search space and assume that

$$f_i(x) : \mathscr{S} \to \mathbb{R}, \quad i = 1, 2, \ldots, k,$$

are k *objective functions* defined over $\mathscr{S}$. Let us assume that,

$$g_j(x) \leqslant 0, \quad j = 1, 2, \ldots, m,$$

are m inequality *constraints*, then the MO problem can be stated as follows: Detect the point:

$$x^* = (x_1^*, x_2^*, \ldots, x_n^*) \in \mathscr{S},$$

that fulfills the constraints and optimizes the following function:

$$F_{nk}(x) = \big(f_1(x), f_2(x), \ldots, f_k(x)\big) : \mathbb{R}^n \to \mathbb{R}^k.$$

The objective functions may be conflicting with each other, thus, it is usually impossible to find the global minimum for all the objectives at the same point. The aim of MO is to provide a set of *Pareto optimal solutions* (points) to the above-mentioned problem.

Specifically, assume that $u = (u_1, u_2, \ldots, u_k)$ and $v = (v_1, v_2, \ldots, v_k)$ are two vectors. Then, u *dominates* v if and only if $u_i \leqslant v_i$, for $i = 1, 2, \ldots, k$, and $u_i < v_i$ for at least one component. This condition is known as *Pareto dominance* and it is used to determine the Pareto optimal solutions. Therefore, a solution x of the MO problem is called *Pareto optimal* if and only if there is not another solution y, such that $F_{nk}(y)$ dominates $F_{nk}(x)$.

The set of all Pareto optimal solutions of an MO problem, denoted by $\mathscr{P}^*$, is called *Pareto optimal set* while the set:

$$\mathscr{P}\mathscr{F}^* = \big\{\big(f_1(x), f_1(x), \ldots, f_k(x)\big) \mid x \in \mathscr{P}^*\big\},$$

is called *Pareto front*. A Pareto front $\mathscr{P}\mathscr{F}^*$ is said to be *convex* if and only if there exists a $w \in \mathscr{P}\mathscr{F}^*$, such that:

$$\lambda\|u\| + (1 - \lambda)\|v\| \geqslant \|w\|, \quad \forall\, u, v \in \mathscr{P}\mathscr{F}^*, \,\forall\, \lambda \in (0, 1),$$

while it is called *concave* if and only if there exists a $w \in \mathscr{P}\mathscr{F}^*$, such that:

$$\lambda\|u\| + (1 - \lambda)\|v\| \leqslant \|w\|, \quad \forall\, u, v \in \mathscr{P}\mathscr{F}^*, \,\forall\, \lambda \in (0, 1).$$

A Pareto front can be convex, concave, or partially convex and/or concave and/or discontinuous. The last three cases exhibit the greatest difficulty for the majority of MO techniques.

Using the MO approach is desirable to detect all the Pareto optimal solutions. On the other hand, the Pareto optimal set may be infinite and since the computation is usually restricted within strict time and space limitations, the main aim of MO is the detection of the largest possible number of Pareto optimal solutions, with the smallest possible deviation from the Pareto front and suitable spread [93].

The *evolutionary algorithms* have the ability to evolve multiple Pareto optimal solutions simultaneously and thus, they are particularly efficient and effective in tackling MO problems. The detected Pareto optimal solutions are stored in memory structures, called *external archives* which, in general, increase the performance of MO approaches. A plethora of well-known and widely applied MO evolutionary approaches have been proposed that are based on different approaches including *niching fitness sharing* and *elitism*, among others [23, 30, 36, 52, 93, 114, 131].

3 Data Preprocessing

The *machine learning (ML) algorithms* aim to automate the process of knowledge extraction from formats that can be easily processed by computer systems. In general, the "quality of the data" could decrease the performance of a learning algorithm. Thus, data preprocessing [40] is an important task in the machine learning pipeline that usually is executed by removing objects and features that contain extraneous and irrelevant information.

Feature Selection The task of detecting and eliminating irrelevant and redundant *features* also known as *attributes* is called *feature selection* (FS) [110]. This task tries to compact the cardinality of the data attributes and to assist the learning algorithms in order to function faster and more efficiently. In general, features can be distinguished as follows:

(a) *Relevant*: Features that contribute an important role for the class and they cannot be assumed by the remaining ones.
(b) *Irrelevant*: Features that do not have any influence on the target class.
(c) *Redundant*: Features that can be replaced by other features.

By eliminating the irrelevant and redundant features, the FS process could assist towards decreasing the training time as well as to simplify the learned models and/or to improve the performance measure of the problem. In general, FS could be considered as a multi-objective task. The main objectives are two: the first one is the maximization of the model's performance while the second one is the minimization of the number of features that will be fed in the learning algorithm. The aforementioned objectives are conflicted and the optimal choice has to be made by considering a balance between the two objectives. Multi-objective FS can acquire a set of non-dominated feature splits in order to meet diverse requirements in real-world applications.

Next, we briefly present various approaches for FS. The efficient and effective *particle swarm optimization* (PSO) method [59, 93] is considered as a metaheuristic approach that attempts to solve an optimization task by maintaining a population of candidate solutions (which are called particles). The members of the swarm are moving around the search space according to a mathematical model that tackles two parameters: the particles' position and velocity. Xue et al. [125] conducted a study on different types of *multi-objective PSO* for FS. The main objective of their work was to create a *PSO-based multi-objective FS scheme* in order to tackle classification problems. Their approach tries to achieve a Pareto front of non-dominated solutions, which will contain a subset of the initial feature space, by simultaneously achieving a more accurate classification performance without using all the available attributes.

Han and Ren [48] proposed a multi-objective technique to improve the performance of FS. They believe that their method could meet different requirements as well as to achieve a trade-off between different conflicting objectives.

Paul and Das [96] proposed an FS and the weighting method supported by an evolutionary multi-objective algorithm on decomposition. The instance attributes are selected and weighted, or scaled and at the same time the data points are displayed to a specific hyper-space. Furthermore, the distances between the data points of the non-identical classes are increased in such a way to facilitate their classification.

Wang et al. [121] presented an algorithm called MECY-SF by using class-dependent redundancy for the FS procedure. Their algorithm exploits genetic search and multi-objective optimization to overcome the limitations of greedy FS algorithms. Furthermore, the fast and elitist multi-objective genetic algorithm named NSGA-II [23] was adopted to solve the multi-objective feature selection problem. Xue et al. [126] gave recently an up-to-date review of the most promising works on evolutionary computation for FS, which provides the contributions of these algorithms.

Cano et al. [11] through their new multi-objective method have succeeded feature extraction and data visualization. Their algorithm is based on Pareto optimal set and is combined with genetic programming. Various classification and visualization measures were assumed as objectives to be optimized by their algorithm.

Das and Das [19] formulated the FS as a *bi-objective optimization problem* of some real-valued weights that correspond to each attribute in the feature space. Therefore, a subset of the weighted attributes is selected as the best subset for subsequent classification of the dataset. The *relevancy* and *redundancy* measures were selected for creating the objective functions.

The FS problem was handled by Hancer et al. [49] through a new *multi-objective artificial bee colony (ABC) algorithm*. Specifically, the authors developed an FS method that will search for a Pareto optimal set of features. They proposed two versions, namely the *binary multi-objective artificial bee colony* named Bin-MOABC version and the corresponding continuous version Num-MOABC. Their algorithm approaches the multi-objective problem through the minimum selection of features that provides the lower classification error in accordance to the original set of features. They tested the proposed algorithm in twelve benchmark datasets and their experimental results show that the Bin-MOABC algorithm exhibits a better classification accuracy and outperforms the other considered methods regarding the dimensionality reduction.

Last but not least, Zheng and Wang [129] proposed an FS method that combines the *joint maximal information entropy* (JMIE), as a measurement metric of a feature subset, and a *binary particle swarm optimization* (BPSO) algorithm for searching the optimal set of features. The authors conducted experiments on five UCI datasets and their experimental results show that the provided technique exhibits a better performance in FS with multiple classes. In addition, their method is more consistent and achieves a better time-efficiency than the BPSO-SVM algorithm.

Instance Selection The *instance selection* or *prototype selection* [25] can be considered as an optimization problem since required the maintenance of mining quality, at first, and secondary the minimization of the sample size. The complexity of the induced solution tends to be increased by the number of the training examples. On the other hand, this may decrease the interpretability of the results. Thus, instance selection is highly recommended in the case of big datasets. Fernández et al. [33] used a multi-objective evolutionary algorithm for searching to obtain the best joint set of both features and instances.

Acampora et al. [1] proposed a multi-objective optimization scheme for the *training set selection (TSS) problem*. The main difference between the provided technique and the evolutionary approaches that had already been developed is the *multi-objective a priori technique*. This means that their method maintains two objectives, namely the *classification accuracy* and the *rate reduction*, unlike all the other evolutionary methods for TSS problem in *support vector machines* (SVM). The authors tested their method using the UCI datasets and the conducted experiments show that the provided algorithm exhibits a better performance on well-known TSS techniques and reinforces the efficiency of SVMs.

Missing Data Imputation The *incomplete* or *corrupted* data values is a common problem [69] in many of the real-life databases. There are many considerations that have to be keep in view in accordance with processing unknown attributes. Determining the origin of the *unknownness* is a major issue. Thus, we lead to the following reasons:

1. The feature is omitted because somehow it was forgotten or for some reason it got lost.
2. For a given object a specific feature value is not applicable.
3. The training dataset collector may not interested in a specific feature value for a given instance.

Lobato et al. [70] presented a multi-objective *genetic algorithm* for data imputation, based on the fast and elitist multi-objective genetic algorithm called NSGA-II [23], which is suitable for mixed (categorical and continuous) attribute datasets and it considers information from incomplete instances and the modelling task. In order to compute the objective function, the following two most common evaluation measures were chosen: (a) the *root mean square error* and (b) the *classification accuracy.*

Discretization The *discretization* assists to the transformation of the space of real-valued attributes to a fixed number of distinct discrete values. A large number of possible feature values could lead to time-consuming machine learning learners. The choice of the number of bins in the discretization process remains an open problem.

Tahan and Asadi [116] proposed an evolutionary approach for the discretization process by using two objectives. The first objective function minimizes the classification error, while the second one minimizes the number of cut points.

Imbalanced Datasets The ideal situation for a *supervised predictor* is to generalize over unknown objects of any class with the same accuracy. In real-life tasks, learners deal with *imbalanced datasets* [13]. This phenomenon leads the learner to be "subjective" towards one class. This can happen when one class is greatly under-represented in the training set in relation to the others. It is associated with training of learning algorithms.

Algorithms in the *inductive machine learning* usually are designed to minimize a predefined metric over a training dataset. Moreover, if any class contains a small amount of examples, in the most of the cases, it can be ignored by the learning algorithms. This is because the cost of performing well on the over-represented class outweighs the cost of doing poorly on the smaller class. Recently, a *convex-hull-based multi-objective genetic programming algorithm* was proposed [128]. This algorithm was applied to binary classification cases and achieved to maximize the convex hull area by minimizing the false positive rate and maximizing the true positive rate simultaneously. The area under the *receiver operating characteristic* (ROC) curve was used as a performance assessment and for the guidance of the search.

Zhao et al. [128] in their attempt to improve the 2D ROC space incorporated the complexity to the objectives. This led to the creation of a 3D objective space (in contrast with the previous 2D ROC space). Li et al. [67] applied swarm optimization on two aspects for re-balancing the imbalanced datasets. One aspect is the search for the appropriate amount of majority instances, while the other one is the estimation of the best control parameters, namely the intensity and the distance of the neighbors of the minority samples in order to be synthesized.

4 Supervised Learning

In machine learning, the *classification* [61] is the paradigm where an algorithm is trained using a training set of correctly identified instances in such a way to produce a model that will be able to correctly identify unseen objects.

Decision Trees The *decision trees* [99, 102] classify examples starting from the root node and afterwards they sort them based on their feature values. In a decision tree each *node* represents a feature of an instance to be classified, while each *branch* represents a value that the node can have.

Zhao [127] proposed a multi-objective genetic programming approach in order to develop a *Pareto optimal decision tree*. This implementation allows the user to select priorities for the conflicting objectives, such as *false negative* versus *false positive*, *sensitivity* versus *specificity*, and *recall* versus *precision*.

Fieldsend [34] used the *particle swarm optimization* (PSO) method [59, 93] in order to train near optimal decision tree using the multi-objective formulation for trading off error rates in each class.

Basgalupp et al. [7] proposed a genetic algorithm for inducing decision trees called LEGAL-Tree. Specifically, they proposed a *lexicographic* approach, where multiple objectives are evaluated in the order of their priority.

Barros et al. [6] provided a taxonomy which groups works that evolve decision trees using evolutionary algorithms. Chikalov et al. [14] created bi-criteria optimization problems for decision trees. The authors considered different cost functions such as *number of nodes*, *depth*, and *average depth*. They design algorithms that are able to determine Pareto optimal points for a given decision table.

Rule Learners The *classification rules* [38, 123] represent each class by the *disjunctive normal form*. The aim is to find the smallest rule-set that is consistent with the training set. Many produced rules are usually a sign that the learning algorithm over-fits the training data.

Dehuri et al. [24] gave an *elitist multi-objective genetic algorithm* (EMOGA) for producing classification rules. They proposed a multi-objective genetic algorithm with a hybrid crossover operator for simultaneously optimizing the objectives of the *comprehensibility*, the *accuracy*, and the *interestingness of rules*.

Pappa and Freitas [87] also successfully produced accurate as well as compact rule models using a *multi-objective grammar-based genetic programming* algorithm.

Srinivasan and Ramakrishnan [115] tackled the problem of discovering rules as a multi-objective optimization problem. They used an approach with three objectives to be optimized. These were metrics such as *accuracy*, *comprehensibility*, and *novelty*.

Rudzinski [104] presented a multi-objective genetic approach in order to produce *interpretability-oriented fuzzy rules* from data. Their proposed approach allows the user to obtain systems with various levels of compromise between *accuracy* and *interpretability*.

Bayesian Classifiers A *Bayesian network* (BN) [51, 124] is a graphical model for probabilistic relationships among the variables. The structure S of a BN is a *directed acyclic graph*. The nodes in S are in one-to-one correspondence with the variables and the arcs represent casual influences among the variables. The lack of possible arcs in S represents conditional independency, while a node (variable) is conditionally independent from its non-descendants given its parents.

Rodriguez and Lozano [101] introduced a structural learning approach of a multi-dimensional Bayesian learner based on the fast and elitist multi-objective genetic algorithm NSGA-II [23].

Panda [86] used the so-called ENORA algorithm which is an FS multi-objective evolutionary algorithm for multi-class classification problems. Specifically, the author estimated the averaged 1-dependence estimators of naive Bayes, through the aforesaid algorithm. The proposed scheme was tested on twenty one real-world datasets and the experimental results show that the implementation of the method is promising in terms of time and accuracy.

Support Vector Machines The *support vector machine* (SVM) [50, 109] is a classification model that is based on the *structured risk minimization theory*. Selecting C, kernel, and γ parameters of SVM is crucial for producing an efficient SVM model. The parameter C of the radial basis function (RBF) kernel SVM compromises misclassification of the training examples contrary to simplicity of the decision surface. A low value of the parameter C causes the decision surface smooth, while a high value of C attempts at classifying all the training examples correctly by providing the model freedom to select more samples as support vectors. The γ parameters can be considered as the inverse of the radius of influence of samples selected by the model as support vectors.

Aydin et al. [5] used a *multi-objective artificial immune algorithm* in order to optimize the kernel as well as the parameters of SVM. Miranda et al. [76] proposed a *hybrid multi-objective architecture* which combines *meta-learning* with *multi-objective particle swarm optimization algorithms* in order to tackle the SVM parameter selection problem.

Gu et al. [45] proposed a *bi-parameter space partition algorithm* for SVMs, which is able to fit all the solutions for every parameter pair. Based on the bi-parameter space partition, they proposed a K-fold cross-validation algorithm for computing the global optimum parameter pairs.

Rosales-Perez et al. [103] used an *evolutionary multi-objective model* and instance selection for SVMs for producing Pareto-based ensemble. Their aims were to minimize the size of the training data and maximize the classification accuracy by the selecting instances.

Neural Networks It is well known that the *perceptrons* [106, 107] are only able to classify linearly separable sets of instances. If the instances are not linearly separable, learning will never find a hyperplane for which all examples are correctly classified. To this end, the *multilayered perceptrons* (artificial neural networks) have been proposed in order to tackle this problem.

Tan et al. [117] used a *modified micro-genetic algorithm optimizer* for twofold, i.e., to select a small number of input features for classification and to improve the accuracy of the neural network model.

Ojha et al. [84] proposed a *multi-objective genetic program* (MOGP) in order to create a *heterogeneous flexible neural tree*, which is a tree-like flexible feed-forward neural network model.

Lazy Learners The *K-nearest neighbor* (*k*-NN) [18, 72] is based on the principle that the instances within a dataset will generally share similar properties. The *k*-NN finds the *k* nearest instances to the testing instance and predicts its class by identifying the most frequent class. Prototype generation is the generation of a small set of instances to replace the initial data, in order to be used by *k*-NN for classification. The main aspects to consider when implementing a prototype generation method are:

(a) the accuracy of a *k*-NN classifier using the prototypes and
(b) the percentage of dataset reduction.

Both factors are in conflict and thus this problem can be naturally handled with multi-objective optimization techniques.

Escalante et al. [31] proposed a *multi-objective evolutionary algorithm for prototype generation*, named MOPG. In addition, Hu and Tan [54] presented a prototype generation using a *multi-objective particle swarm optimization for k-NN classifier*.

Ensembles The selection of a single algorithm in order to produce a reliable classification model is not an easy task. A simple approach could be the estimation of the accuracy of the candidate algorithms on a problem and then the selection of the best performer. The idea of combining classifiers [26, 27] is proposed as a direction for increasing the classification accuracy in real-world problems. In this case, the objective is to use the strengths of one model to complement the weaknesses of the other. In general, the multi-objective evolutionary algorithms for the construction of classifier ensembles is an interesting area of study and research.

Chandra and Yao [12] presented an *ensemble learning algorithm*, which is named DIVACE (DIVerse and ACcurate Ensemble learning algorithm). This algorithm tries to find a trade-off between diversity and accuracy by treating these two objectives explicitly separately. Three other approaches for the *Pareto-based multi-objective ensemble generation* approach are compared and discussed in [57].

Bhowan et al. [9] proposed a *multi-objective genetic programming* method in order to evolve accurate and diverse classifiers with acceptable accuracy both on the minority and majority of class. Furthermore, Bhowan et al. [10] presented another similar approach in order to evolve ensembles by using genetic programming for imbalanced data.

Nguyen et al. [83] used a genetic algorithm approach that focuses on the following three objectives:

1. The count of correct classified instances,
2. The count of selected attributes, and
3. The count of selected classifiers.

Gu et al. [46] presented a survey on multi-objective ensemble generation methods, including the diversity measures, member generation, as well as the selection and integration techniques.

Nag and Pal [80] presented an integrated algorithm for simultaneous attribute selection and inducing diverse learners using a steady state multi-objective genetic programming, which minimizes the following three objectives:

(a) False positives predictions,
(b) False negatives predictions, and
(c) The count of leaf nodes in the decision tree.

Albukhanajer et al. [3] propose classifier ensembles that use multiple Pareto image features for invariant image identification.

Last but not least, very recently, Pourtaheri et al. [98] developed two multi-objective heuristic ensemble classifiers by combining the *multi-objective inclined planes optimization* algorithm and the *multi-objective particle swarm optimization* (MOPSO) algorithm.

5 Unsupervised Learning

Clustering The *cluster analysis* [88] is a process that is very useful for the exploration of a collection of data. As it is implied by the term "cluster," through this process, elements or features with an inter-relationship are detected, which can lead to the homogeneous clustering of data. It can be either *supervised* or *unsupervised* and the major difference between this process and the classification process is that the first one does not use labels to assist in the categorization of the data in order to create a cluster structure. Furthermore, whether an item belongs to a particular cluster is determined through an intra-connectivity measurement. If this measure is high, it means that the clusters are "compact" and the data of the same group are highly dependent on each other. On the other hand, the inter-connectivity measurement is a criterion that declares the independence between the clusters. Thus, if the inter-connectivity is low, it means that the individual clusters are largely independent to each other. More details about *multi-objective evolutionary clustering algorithms*, the reader can reach in [29, 79].

The *mathematical programming* [32] has an important contribution to the issue of cluster analysis. The direct connection of the two areas can be easily understood, since it is required the minimum number of clusters to which the original dataset can be grouped. Thus, this approach can be considered as an optimization problem, with specific features and constraints. An important issue is also the appropriate selection of solutions, since from a set of feasible, "good" solutions, the best solutions are those of interest [89].

Luo et al. [71] proposed a method for modelling spectral clustering and through specific operators they selected a set of good individuals at the optimization process. Furthermore, the authors through the *ratio cut criterion* selected a trade-off solution from the Pareto set. Finally, the various problems that have been analyzed for supervised and unsupervised classification tasks contributed to the creation of semi-supervised clustering techniques. With a small amount of labelled data and the data distribution, Alok et al. [4] proposed a *semi-supervised clustering method* by using the multi-objective optimization framework.

Wang et al. [122], recently, through an *evolutionary multi-objective* (EMO) algorithm tackled a very difficult and timeless challenge for a clustering method problem, namely the *"determination of the number of clusters."* To this end, the authors proposed a scheme that uses an EMO algorithm, specifically the rapid elitist multi-objective genetic algorithm named NSGA-II, in order to select the non-dominated solutions. The process that follows includes a validity index for selecting the optimal clustering result. The authors tested their model on three datasets and their experimental results show that the *EMO-k-clustering method* is effective and by executing only a single run it is able to obtain all the clustering results for different values of the parameter k.

Last but not least, very recently, Nayak et al. [81] proposed the *elitism-based multi-objective differential evolution* (EMODE) algorithm for automatic clustering. Their work handles complex datasets using three objectives. The authors conducted experiments on ten datasets and the results show that their approach provides an alternative solution for data clustering in many different areas.

Association Rules The *association rule mining* (ARM) [111] has as a primary goal the discovery of associations rules between data of a given database. The first goal of this procedure is to come up with the data that have the greatest appearance in the database. Then, the appropriate association rules are created for the whole dataset by using the feature values that their appearance exceed a certain predetermined threshold.

Minaei-Bidgoli et al. [75] proposed a *multi-objective genetic algorithm* for mining association rules from numerical variables. It is known that well-known and widely used models that handle the association rule mining process cannot be applied to datasets which consist of numerical data. For this reason, it is necessary the preprocessing of the data and in particular the discretization process. Minaei-Bidgoli et al. [75], using three measures, namely: *confidence, interestingness*, and *comprehensibility*, defined three different objective functions for their approach and extracted the best association rules through Pareto optimality.

Beiranvand et al. [8] proposed a *multi-objective particle swarm optimization model* named MOPAR for mining numerical association rules in only one single step without a priori discretization. The authors conducted experiments and the results show that their approach extracts reliable, comprehensible, and interesting numerical association rules.

Martin et al. [73] proposed a *multi-objective evolutionary model* named QAR-CIP-NSGA-II which extents the well-known elitist multi-objective genetic algorithm named NSGA-II [23]. Their method performs an evolutionary learning and a selection condition for each association rule. Furthermore, their approach maximizes two of the three objective functions that Minaei-Bidgoli et al. considered. In addition, their approach maximizes the performance of the objective functions for mining a set of quantitative association rules with enough interpretability as well as accurate results.

6   A Few of the Most Recent Applications

The various applications that have been provided over the last years show the importance of the *multi-objective evolutionary optimization algorithms* (MOEOA). A few of the most recent and very interesting applications regarding MOEOA are the following ones.

Mason et al. [74] developed an artificial neural network that has been trained through a *differential evolution* (DE) algorithm. Their proposed neural network has the ability to handle multi-objective optimization problems using properly an approximation function. Specifically, the proposed approach uses a single objective global optimizer (the DE algorithm) in order to evolve the neural network. In other words, the so-called MONNDE algorithm is capable to provide further *Pareto fronts* without any further optimization effort. The authors applied the MONNDE algorithm to the well-known *dynamic economic emission dispatch* problem and through the experiments that they conducted, they show that the performance of their algorithm is equally optimal in comparison with other well-known and widely used algorithms. Furthermore, they show that it is more efficient to optimize the topology of the neural network dynamically with an online way, instead of to optimize the weights of the neural network.

Rao et al. [100] proposed an alternative *classifier for disease diagnosis*. Specifically, the proposed scheme includes a sequential minimal optimization, the SVM classifier, and three evolutionary algorithms for the evolution of the parameters. Moreover, the authors presented a new technique, which is named *cuboids elephant herding optimization* (CEHO). Their approach is applied to seventeen medical datasets and the experimental results show that the proposed technique exhibits a very good performance for all the tested datasets.

Sabar et al. [105] considered the configuration of a SVM as a bi-objective optimization problem. The accuracy of the model was the first objective while the other one was the complexity of the model. The authors proposed a novel hyper-heuristic framework for the optimization of the above-mentioned two conflicting objectives. The developed approach tested on two *cyber security problems* and the experimental results show that their proposed scheme is very effective and efficient.

7   Synopsis and Discussion

In general, the subject of the *multi-objective evolutionary optimization algorithms* (MOEOA) is related to an interesting concept with many different aspects and a crucial role, not only in machine learning, but also in many other scientific fields. This is evident, since in nowadays, the necessity of handling conflicting performance objectives appears in many scientific fields. The plethora of papers written regarding MOEOA show in an emphatic way that the scientific community has a great concern about this subject.

The very first evolutionary approaches to solve multi-objective optimization problems and especially the particle swarm optimization and differential evolution algorithms appeared very promising [91–95]. It is worth mentioning that the vector evaluated particle swarm optimization and the vector evaluated differential evolution [91, 95] remain the basis of current research in *multi-objective optimization, many-objective optimization*, and *dynamic multi-objective optimization*. Furthermore, the multi-objective optimization has led to better performing machine learning models in contrast to the traditional single objective ones.

The importance of multi-objective evolutionary algorithms is apparent not only from the plethora of papers that have been presented by the scientific community, but also from a huge amount of various applications that have been presented over the last decades such as engineering [43], industry [64], economy [16, 66], and many others [2, 28, 39, 41, 42, 53, 65, 68, 97]. The reader could also reach more details about the variety of the problems and the amount of applications in [60] and [120].

This paper describes how multi-objective evolutionary optimization algorithms have been used in the field of machine learning, in relative detail. It should be noted that the list of the references in the current work does not provide a complete list of the papers corresponding to this subject. The main aim was to provide a survey of the basic ideas, rather than a simple list, of all the research papers that have been discussed or have been used these ideas. Nevertheless, it is hoped that the mentioned references will cover the most important theoretical issues and will give guidelines to the main branches of literature regarding such techniques and schemes, guiding the interested reader to up-to-date research directions.

Acknowledgements S.-A. N. Alexandropoulos is supported by Greece and the European Union (European Social Fund-ESF) through the Operational Programme "Human Resources Development, Education and Lifelong Learning" in the context of the project "Strengthening Human Resources Research Potential via Doctorate Research" (MIS-5000432), implemented by the State Scholarships Foundation (IKY).

References

1. Acampora, G., Herrera, F., Tortora, G., Vitiello, A.: A multi-objective evolutionary approach to training set selection for support vector machine. Knowl. Based Syst. **147**, 94–108 (2018)
2. Ahmed, F., Deb, K., Jindal, A.: Multi-objective optimization and decision making approaches to cricket team selection. Appl. Soft Comput. **13**(1), 402–414 (2013)
3. Albukhanajer, W.A., Jin, Y., Briffa, J.A.: Classifier ensembles for image identification using multi-objective Pareto features. Neurocomputing **238**, 316–327 (2017). https://doi.org/10.1016/j.neucom.2017.01.067
4. Alok, A.K., Saha, S., Ekbal, A.: A new semi-supervised clustering technique using multi-objective optimization. Appl. Intell. **43**(3), 633–661 (2015). https://doi.org/10.1007/s10489-015-0656-z
5. Aydin, I., Karakose, M., Akin, E.: A multi-objective artificial immune algorithm for parameter optimization in support vector machine. Appl. Soft Comput. **11**(1), 120–129 (2011). https://doi.org/10.1016/j.asoc.2009.11.003

6. Barros, R.C., Basgalupp, M.P., De Carvalho, A.C., Freitas, A.A.: A survey of evolutionary algorithms for decision-tree induction. IEEE Trans. Syst. Man Cybern. Part C Appl. Rev. **42**(3), 291–312 (2012)

7. Basgalupp, M.P., Carvalho, A.C.D., Barros, R.C., Ruiz, D.D., Freitas, A.A.: Lexicographic multi-objective evolutionary induction of decision trees. Int. J. Bio-Inspired Comput. **1**(1/2), 105 (2009). https://doi.org/10.1504/ijbic.2009.022779

8. Beiranvand, V., Mobasher-Kashani, M., Bakar, A.A.: Multi-objective PSO algorithm for mining numerical association rules without a priori discretization. Expert Syst. Appl. **41**(9), 4259–4273 (2014). https://doi.org/10.1016/j.eswa.2013.12.043

9. Bhowan, U., Johnston, M., Zhang, M., Yao, X.: Evolving diverse ensembles using genetic programming for classification with unbalanced data. IEEE Trans. Evol. Comput. **17**(3), 368–386 (2013). https://doi.org/10.1109/tevc.2012.2199119

10. Bhowan, U., Johnston, M., Zhang, M., Yao, X.: Reusing genetic programming for ensemble selection in classification of unbalanced data. IEEE Trans. Evol. Comput. **18**(6), 893–908 (2014). https://doi.org/10.1109/tevc.2013.2293393

11. Cano, A., Ventura, S., Cios, K.J.: Multi-objective genetic programming for feature extraction and data visualization. Soft. Comput. **21**(8), 2069–2089 (2017). https://doi.org/10.1007/s00500-015-1907-y

12. Chandra, A., Yao, X.: Ensemble learning using multi-objective evolutionary algorithms. J. Math. Model. Algorithms **5**(4), 417–445 (2006). https://doi.org/10.1007/s10852-005-9020-3

13. Chawla, N.V.: Data mining for imbalanced datasets: an overview. In: Data Mining and Knowledge Discovery Handbook, pp. 875–886. Springer, Boston (2009)

14. Chikalov, I., Hussain, S., Moshkov, M.: Bi-criteria optimization of decision trees with applications to data analysis. Eur. J. Oper. Res. **266**(2), 689–701 (2018). https://doi.org/10.1016/j.ejor.2017.10.021

15. Coello Coello, C.A.: A comprehensive survey of evolutionary-based multiobjective optimization techniques. Knowl. Inf. Syst. **1**(3), 269–308 (1999)

16. Coello Coello, C.A.: Evolutionary multi-objective optimization and its use in finance. In: Rennard, J.-P. (ed.) Handbook of Research on Nature Inspired Computing for Economy and Management. Idea Group Publishing, Hershey (2006)

17. Costa, L., Oliveira, P.: Dimension reduction in multiobjective optimization. Proc. Appl. Math. Mech. **7**(1), 2060047–2060048 (2007)

18. Cunningham, P., Delany, S.J.: k-nearest neighbour classifiers. Mult. Classif. Syst. **34**, 1–17 (2007)

19. Das, A., Das, S.: Feature weighting and selection with a Pareto-optimal trade-off between relevancy and redundancy. Pattern Recogn. Lett. **88**, 12–19 (2017). https://doi.org/10.1016/j.patrec.2017.01.004

20. Deb, K.: Multi-Objective Optimization using Evolutionary Algorithms. Wiley, Chichester (2001)

21. Deb, K.: Multi-objective optimization using evolutionary algorithms: an introduction. In: Wang, L.-H., Ng, A.H.C., Deb, K. (eds.) Multi-Objective Evolutionary Optimisation for Product Design and Manufacturing, pp. 401–430. Springer, London (2011)

22. Deb, K.: Multi-objective optimization. In: Burke, E.K., Kendall, G. (eds.) Search Methodologies, pp. 403–449. Springer, Boston (2014)

23. Deb, K., Pratap, A., Agarwal, S., Meyarivan, T.: A fast and elitist multiobjective genetic algorithm: NSGA-II. IEEE Trans. Evol. Comput. **6**(2), 182–197 (2002)

24. Dehuri, S., Patnaik, S., Ghosh, A., Mall, R.: Application of elitist multi-objective genetic algorithm for classification rule generation. Appl. Soft Comput. **8**(1), 477–487 (2008). https://doi.org/10.1016/j.asoc.2007.02.009

25. Derrac, J., García, S., Herrera, F.: A survey on evolutionary instance selection and generation. Int. J. Appl. Metaheuristic Comput. **1**, 60–92 (2010)

26. Dietterich, T.G.: Ensemble methods in machine learning. In: Multiple Classifier Systems, MCS 2000. Lecture Notes in Computer Science, vol. 1857, pp. 1–15. Springer, Berlin (2000)

27. Díez-Pastor, J.F., Rodríguez, J.J., García-Osorio, C.I., Kuncheva, L.I.: Diversity techniques improve the performance of the best imbalance learning ensembles. Inf. Sci. **325**, 98–117 (2015)
28. Domingo-Perez, F., Lazaro-Galilea, J.L., Wieser, A., Martin-Gorostiza, E., Salido-Monzu, D., de la Llana, A.: Sensor placement determination for range-difference positioning using evolutionary multi-objective optimization. Expert. Syst. Appl. **47**, 95–105 (2016)
29. Dutta, D., Dutta, P., Sil, J.: Simultaneous feature selection and clustering with mixed features by multi objective genetic algorithm. Int. J. Hybrid Intell. Syst. **11**(1), 41–54 (2014). https://doi.org/10.3233/HIS-130182
30. Erickson, M., Mayer, A., Horn, J.: The niched Pareto genetic algorithm 2 applied to the design of groundwater remediation systems. Lect. Notes Comput. Sci. **1993**, 681–695 (2001)
31. Escalante, H.J., Marin-Castro, M., Morales-Reyes, A., Graff, M., Rosales-Pérez, A., y Gómez, M.M., Reyes, C.A., Gonzalez, J.A.: MOPG: a multi-objective evolutionary algorithm for prototype generation. Pattern. Anal. Applic. **20**(1), 33–47 (2015). https://doi.org/10.1007/s10044--015-0454--6
32. Fan, Y.J., Iyigun, C., Chaovalitwongse, W.A.: Recent advances in mathematical programming for classification and cluster analysis. In: CRM Proceedings & Lecture Notes, vol. 45, pp. 67–93 (2008)
33. Fernández, A., Carmona, C.J., del Jesus, M.J., Herrera, F.: A Pareto-based ensemble with feature and instance selection for learning from multi-class imbalanced datasets. Int. J. Neural Syst. **27**(06), 1750028 (2017). https://doi.org/10.1142/s0129065717500289
34. Fieldsend, J.E.: Optimizing decision trees using multi-objective particle swarm optimization. In: Studies in Computational Intelligence, pp. 93–114. Springer, Berlin (2009). https://doi.org/10.1007/978-3-642-03625-5_5
35. Floudas, C.A., Pardalos, P.M.: Encyclopedia of Optimization, vol. 2. Springer, Berlin (2001)
36. Fonseca, C.M., Fleming, P.J.: Genetic algorithms for multiobjective optimization: formulation, discussion and generalization. In: Proceedings of the Fifth International Conference on Genetic Algorithms (ICGA 1993), Urbana-Champaign, IL, USA, pp. 416–423 (1993)
37. Freitas, A.A.: A critical review of multi-objective optimization in data mining. ACM SIGKDD Explor. Newsl. **6**(2), 77 (2004). https://doi.org/10.1145/1046456.1046467
38. Fürnkranz, J.: Separate-and-conquer rule learning. Artif. Intell. Rev. **13**(1), 3–54 (1999)
39. Ganesan, T., Elamvazuthi, I., Vasant, P.: Multiobjective design optimization of a nano-CMOS voltage-controlled oscillator using game theoretic-differential evolution. Appl. Soft Comput. **32**, 293–299 (2015)
40. García, S., Luengo, J., Herrera, F.: Data Preprocessing in Data Mining. Springer, Cham (2015)
41. Giannopoulos, N., Nearchou, A.C.: Bi-criteria scheduling against restrictive common due dates using a multi-objective differential evolution algorithm. IMA J. Manag. Math. **29**(1), 119–136 (2018)
42. Giannopoulos, N., Moulianitis, V.C., Nearchou, A.C.: Multi-objective optimization with fuzzy measures and its application to flow-shop scheduling. Eng. Appl. Artif. Intell. **25**(7), 1381–1394 (2012)
43. Gong, W., Cai, Z., Zhu, L.: An efficient multiobjective differential evolution algorithm for engineering design. Struct. Multidiscip. Optim. **38**(2), 137–157 (2009)
44. Graves, A., Mohamed, A.R., Hinton, G.: Speech recognition with deep recurrent neural networks. In: IEEE International Conference on Acoustics, Speech and Signal Processing (ICASSP), 2013, pp. 6645–6649. IEEE, Piscataway (2013)
45. Gu, B., Sheng, V.S., Li, S.: Bi-parameter space partition for cost-sensitive SVM. In: Proceedings of the 24th International Conference on Artificial Intelligence, IJCAI'15, pp. 3532–3539. AAAI Press, Washington (2015). http://dl.acm.org/citation.cfm?id=2832581.2832741
46. Gu, S., Cheng, R., Jin, Y.: Multi-objective ensemble generation. Wiley Interdiscip. Rev. Data Min. Knowl. Disc. **5**(5), 234–245 (2015). https://doi.org/10.1002/widm.1158
47. Hahsler, M., Grün, B., Hornik, K.: A computational environment for mining association rules and frequent item sets. J. Stat. Softw. **14**, 1–25 (2005)

48. Han, M., Ren, W.: Global mutual information-based feature selection approach using single-objective and multi-objective optimization. Neurocomputing **168**, 47–54 (2015). https://doi.org/10.1016/j.neucom.2015.06.016
49. Hancer, E., Xue, B., Zhang, M., Karaboga, D., Akay, B.: Pareto front feature selection based on artificial bee colony optimization. Inf. Sci. **422**, 462–479 (2018)
50. Hearst, M.A., Dumais, S.T., Osuna, E., Platt, J., Scholkopf, B.: Support vector machines. IEEE Intell. Syst. Appl. **13**(4), 18–28 (1998)
51. Heckerman, D., Geiger, D., Chickering, D.M.: Learning Bayesian networks: the combination of knowledge and statistical data. Mach. Learn. **20**(3), 197–243 (1995)
52. Horn, J., Nafpliotis, N., Goldberg, D.E.: A niched Pareto genetic algorithm for multiobjective optimization. In: Proceedings of the first IEEE International Conference on Evolutionary Computation (ICEC 1994), Orlando, FL, USA, pp. 82–87 (1994)
53. Hruschka, E.R., Campello, R.J., Freitas, A.A., de Carvalho, A.: A survey of evolutionary algorithms for clustering. IEEE Trans. Syst. Man Cybern. Part C Appl. Rev. **39**(2), 133–155 (2009)
54. Hu, W., Tan, Y.: Prototype generation using multiobjective particle swarm optimization for nearest neighbor classification. IEEE Trans. Cybern. **46**(12), 2719–2731 (2016). https;//doi.org/10.1109/tcyb.2015.2487318
55. Huang, G.B., Zhou, H., Ding, X., Zhang, R.: Extreme learning machine for regression and multiclass classification. IEEE Trans. Syst. Man Cybern. B Cybern. **42**(2), 513–529 (2012)
56. Hwang, C.L., Masud, A.S.M.: Multiple Objective Decision Making Methods and Applications: A State-of-the-Art Survey, vol. 164. Springer, Berlin (2012)
57. Jin, Y., Sendhoff, B.: Pareto-based multiobjective machine learning: an overview and case studies. IEEE Trans. Syst. Man Cybern. Part C Appl. Rev. **38**(3), 397–415 (2008). https://doi.org/10.1109/tsmcc.2008.919172
58. Jones, D.F., Mirrazavi, S.K., Tamiz, M.: Multi-objective meta-heuristics: an overview of the current state-of-the-art. Eur. J. Oper. Res. **137**(1), 1–9 (2002)
59. Kennedy, J., Eberhart, R.: Particle swarm optimization. In: Proceedings of the IEEE International Conference on Neural Networks, vol. IV, pp. 1942–1948. IEEE, Piscataway (1995)
60. Khor, E.F., Tan, K.C., Lee, T.H., Goh, C.K.: A study on distribution preservation mechanism in evolutionary multi-objective optimization. Artif. Intell. Rev. **23**(1), 31–33 (2005)
61. Kodratoff, Y.: Introduction to Machine Learning. Morgan Kaufmann, Burlington (2014)
62. Kolari, P., Java, A., Finin, T., Oates, T., Joshi, A.: Detecting spam blogs: a machine learning approach. In: AAAI, vol. 6, pp. 1351–1356 (2006)
63. Kotsiantis, S.B., Zaharakis, I.D., Pintelas, P.E.: Machine learning: a review of classification and combining techniques. Artif. Intell. Rev. **26**(3), 159–190 (2006)
64. Lei, D.: Multi-objective production scheduling: a survey. Int. J. Adv. Manuf. Technol. **43**(9–10), 926–938 (2009)
65. Li, X.: A real-coded predator-prey genetic algorithm for multiobjective optimization. In: International Conference on Evolutionary Multi-Criterion Optimization, pp. 207–221. Springer, Berlin (2003)
66. Li, J., Taiwo, S.: Enhancing financial decision making using multi-objective financial genetic programming. In: IEEE Congress on Evolutionary Computation, 2006. CEC 2006, pp. 2171–2178. IEEE, Piscataway (2006)
67. Li, J., Fong, S., Wong, R.K., Chu, V.W.: Adaptive multi-objective swarm fusion for imbalanced data classification. Inf. Fusion **39**, 1–24 (2018). https://doi.org/10.1016/j.inffus.2017.03.007
68. Lim, D., Ong, Y.S., Jin, Y., Sendhoff, B., Lee, B.S.: Inverse multi-objective robust evolutionary design. Genet. Program Evolvable Mach. **7**(4), 383–404 (2006)
69. Liu, Z.g., Pan, Q., Dezert, J., Martin, A.: Adaptive imputation of missing values for incomplete pattern classification. Pattern Recogn. **52**, 85–95 (2016)
70. Lobato, F., Sales, C., Araujo, I., Tadaiesky, V., Dias, L., Ramos, L., Santana, A.: Multi-objective genetic algorithm for missing data imputation. Pattern Recogn. Lett. **68**, 126–131 (2015). https://doi.org/10.1016/j.patrec.2015.08.023

71. Luo, J., Jiao, L., Lozano, J.A.: A sparse spectral clustering framework via multiobjective evolutionary algorithm. IEEE Trans. Evol. Comput. **20**(3), 418–433 (2016). https://doi.org/10.1109/tevc.2015.2476359

72. Malhotra, S., Bali, V., Paliwal, K.: Genetic programming and k-nearest neighbour classifier based intrusion detection model. In: 7th International Conference on Cloud Computing, Data Science and Engineering-Confluence, 2017, pp. 42–46. IEEE, Piscataway (2017)

73. Martín, D., Rosete, A., Alcalá-Fdez, J., Herrera, F.: QAR-CIP-NSGA-II: A new multi-objective evolutionary algorithm to mine quantitative association rules. Inf. Sci. **258**, 1–28 (2014). https://doi.org/10.1016/j.ins.2013.09.009

74. Mason, K., Duggan, J., Howley, E.: A multi-objective neural network trained with differential evolution for dynamic economic emission dispatch. Int. J. Electr. Power Energy Syst. **100**, 201–221 (2018)

75. Minaei-Bidgoli, B., Barmaki, R., Nasiri, M.: Mining numerical association rules via multi-objective genetic algorithms. Inf. Sci. **233**, 15–24 (2013). https://doi.org/10.1016/j.ins.2013.01.028

76. Miranda, P.B., Prudêncio, R.B., de Carvalho, A.P., Soares, C.: A hybrid meta-learning architecture for multi-objective optimization of SVM parameters. Neurocomputing **143**, 27–43 (2014). https://doi.org/10.1016/j.neucom.2014.06.026

77. Mukhopadhyay, A., Maulik, U., Bandyopadhyay, S., Coello Coello, C.A.: A survey of multiobjective evolutionary algorithms for data mining: Part I. IEEE Trans. Evol. Comput. **18**(1), 4–19 (2014). https://doi.org/10.1109/tevc.2013.2290086

78. Mukhopadhyay, A., Maulik, U., Bandyopadhyay, S., Coello Coello, C.A.: Survey of multi-objective evolutionary algorithms for data mining: Part II. IEEE Trans. Evol. Comput. **18**(1), 20–35 (2014). https://doi.org/10.1109/tevc.2013.2290082

79. Mukhopadhyay, A., Maulik, U., Bandyopadhyay, S.: A survey of multiobjective evolutionary clustering. ACM Comput. Surv. **47**(4), 1–46 (2015). https://doi.org/10.1145/2742642

80. Nag, K., Pal, N.R.: A multiobjective genetic programming-based ensemble for simultaneous feature selection and classification. IEEE Trans. Cybern. **46**(2), 499–510 (2016). https://doi.org/10.1109/tcyb.2015.2404806

81. Nayak, S.K., Rout, P.K., Jagadev, A.K.: Automatic clustering by elitism-based multi-objective differential evolution. Int. J. Manag. Decis. Mak. **17**(1), 50–74 (2018)

82. Newman, M.E.: Power laws, Pareto distributions and Zipf's law. Contemp. Phys. **46**(5), 323–351 (2005)

83. Nguyen, T.T., Liew, A.W.C., Pham, X.C., Nguyen, M.P.: Optimization of ensemble classifier system based on multiple objectives genetic algorithm. In: 2014 International Conference on Machine Learning and Cybernetics. IEEE, Piscataway (2014). https://doi.org/10.1109/icmlc.2014.7009090

84. Ojha, V.K., Abraham, A., Snášel, V.: Ensemble of heterogeneous flexible neural trees using multiobjective genetic programming. Appl. Soft Comput. **52**, 909–924 (2017). https://doi.org/10.1016/j.asoc.2016.09.035

85. Oza, N.C., Tumer, K.: Classifier ensembles: select real-world applications. Inf. Fusion **9**(1), 4–20 (2008)

86. Panda, M.: Combining multi-objective evolutionary algorithm with averaged one-dependence estimators for big data analytics. Int. J. Comput. Intell. Stud. **7**(1), 1–18 (2018)

87. Pappa, G.L., Freitas, A.A.: Evolving rule induction algorithms with multi-objective grammar-based genetic programming. Knowl. Inf. Syst. **19**(3), 283–309 (2008). https://doi.org/10.1007/s10115--008-0171--1

88. Pardalos, P.M., Hansen, P.: Data Mining and Mathematical Programming, vol. 45. American Mathematical Society, Providence (2008)

89. Pardalos, P.M., Žilinskas, A., Žilinskas, J.: Multi-objective branch and bound. In: Non-convex Multi-Objective Optimization, pp. 45–56. Springer, Cham (2017)

90. Pareto, V.: Manuale di economica politica, societa editrice libraria. In: Manual of Political Economy, vol. 1971. A.M. Kelley, New York (1906)

91. Parsopoulos, K.E., Vrahatis, M.N.: Particle swarm optimization method in multiobjective problems. In: Proceedings of the ACM Symposium on Applied Computing (SAC 2002), pp. 603–607. ACM, New York (2002)

92. Parsopoulos, K.E., Vrahatis, M.N.: Multi-objective particle swarm optimization approaches. In: Bui, L.T., Alam, S. (eds.) Multi-Objective Optimization in Computational Intelligence: Theory and Practice, Chapter 2, pp. 20–42. IGI Global, Hershey (2008)

93. Parsopoulos, K.E., Vrahatis, M.N.: Particle Swarm Optimization and Intelligence: Advances and Applications. Information Science Publishing (IGI Global), Hershey (2010)

94. Parsopoulos, K.E., Tasoulis, D.K., Vrahatis, M.N.: Multiobjective optimization using parallel vector evaluated particle swarm optimization. In: Hamza, M.H. (ed.) Artificial Intelligence and Applications, vol. 2, pp.823–828. ACTA Press, Anaheim (2004)

95. Parsopoulos, K.E., Tasoulis, D.K., Pavlidis, N.G., Plagianakos, V.P., Vrahatis, M.N.: Vector evaluated differential evolution for multiobjective optimization. In: Proceedings of the IEEE Congress on Evolutionary Computation (CEC 2004), vol. 1, pp. 204–211. IEEE, Piscataway (2004)

96. Paul, S., Das, S.: Simultaneous feature selection and weighting – an evolutionary multi-objective optimization approach. Pattern Recogn. Lett. **65**, 51–59 (2015). https://doi.org/10.1016/j.patrec.2015.07.007

97. Pena-Reyes, C.A., Sipper, M.: Evolutionary computation in medicine: an overview. Artif. Intell. Med. **19**(1), 1–23 (2000)

98. Pourtaheri, Z.K., Zahiri, S.H., Razavi, S.M.: Stability investigation of multi-objective heuristic ensemble classifiers. Int. J. Mach. Learn. Cybern. 1–13 (2018). https://doi.org/10.1007/s13042-018-0789-6

99. Quinlan, J.R.: Induction of decision trees. Mach. Learn. **1**(1), 81–106 (1986)

100. Rao, N.M., Kannan, K., Gao, X.-Z., Roy, D.S.: Novel classifiers for intelligent disease diagnosis with multi-objective parameter evolution. Comput. Electr. Eng. **67**, 483–496. https://doi.org/10.1016/j.compeleceng.2018.01.039. Available online 13 February 2018 (2018)

101. Rodríguez, J.D., Lozano, J.A.: Multi-objective learning of multi-dimensional Bayesian classifiers. In: 2008 Eighth International Conference on Hybrid Intelligent Systems. IEEE, Piscataway (2008). https://doi.org/10.1109/his.2008.143

102. Rokach, L., Maimon, O.: Data Mining with Decision Trees: Theory and Applications. World Scientific, Singapore (2008)

103. Rosales-Perez, A., Garcia, S., Gonzalez, J.A., Coello Coello, C.A., Herrera, F.: An evolutionary multiobjective model and instance selection for support vector machines with Pareto-based ensembles. IEEE Trans. Evol. Comput. **21**(6), 863–877 (2017). https://doi.org/10.1109/tevc.2017.2688863

104. Rudziński, F.: A multi-objective genetic optimization of interpretability-oriented fuzzy rule-based classifiers. Appl. Soft Comput. **38**, 118–133 (2016). https://doi.org/10.1016/j.asoc.2015.09.038

105. Sabar, N.R., Yi, X., Song, A.: A bi-objective hyper-heuristic support vector machines for big data cyber-security. IEEE Access **6**, 10421–10431 (2018)

106. Samarasinghe, S.: Neural Networks for Applied Sciences and Engineering: From Fundamentals to Complex Pattern Recognition. CRC Press, Boca Raton (2016)

107. Schalkoff, R.J.: Artificial neural networks, vol. 1. McGraw-Hill, New York (1997)

108. Schmidhuber, J.: Deep learning in neural networks: an overview. Neural Netw. **61**, 85–117 (2015)

109. Shen, C., Wang, X., Yu, D.: Feature weighting of support vector machines based on derivative saliency analysis and its application to financial data mining. Int. J. Adv. Comput. Technol. **4**(1), 199–206 (2012)

110. Shu, W., Shen, H.: Multi-criteria feature selection on cost-sensitive data with missing values. Pattern Recogn. **51**, 268–280 (2016)

111. Soysal, Ö.M.: Association rule mining with mostly associated sequential patterns. Expert Syst. Appl. **42**(5), 2582–2592 (2015)

112. Spall, J.C.: Introduction to Stochastic Search and Optimization: Estimation, Simulation, and Control, vol. 65. Wiley, Hoboken (2005)
113. Sra, S., Nowozin, S., Wright, S.J.: Optimization for Machine Learning. MIT Press, Cambridge (2012)
114. Srinivas, N., Deb, K.: Multiobjective optimization using nondominated sorting in genetic algorithms. Evol. Comput. **2**(3), 221–248 (1994). https://doi.org/10.1162/evco.1994.2.3.221
115. Srinivasan, S., Ramakrishnan, S.: Evolutionary multi objective optimization for rule mining: a review. Artif. Intell. Rev. **36**(3), 205–248 (2011). https://doi.org/10.1007/s10462-011-9212-3
116. Tahan, M.H., Asadi, S.: MEMOD: a novel multivariate evolutionary multi-objective discretization. Soft Comput. **22**, 301–323 (2017). https://doi.org/10.1007/s00500-016-2475-5
117. Tan, C.J., Lim, C.P., Cheah, Y.: A multi-objective evolutionary algorithm-based ensemble optimizer for feature selection and classification with neural network models. Neurocomputing **125**, 217–228 (2014). https://doi.org/10.1016/j.neucom.2012.12.057
118. Veldhuizen, D.A.V., Lamont, G.B.: Multiobjective evolutionary algorithms: analyzing the state-of-the-art. Evol. Comput. **8**(2), 125–147 (2000)
119. Viola, P., Jones, M.J.: Robust real-time face detection. Int. J. Comput. Vis. **57**(2), 137–154 (2004)
120. von Lücken, C., Barán, B., Brizuela, C.: A survey on multi-objective evolutionary algorithms for many-objective problems. Comput. Optim. Appl. **58**(3), 707–756 (2014)
121. Wang, Z., Li, M., Li, J.: A multi-objective evolutionary algorithm for feature selection based on mutual information with a new redundancy measure. Inf. Sci. **307**, 73–88 (2015). https://doi.org/10.1016/j.ins.2015.02.031
122. Wang, R., Lai, S., Wu, G., Xing, L., Wang, L., Ishibuchi, H.: Multi-clustering via evolutionary multi-objective optimization. Inf. Sci. **450**, 128–410 (2018)
123. Witten, I.H., Frank, E., Hall, M.A., Pal, C.J.: Data Mining: Practical Machine Learning Tools and Techniques. Morgan Kaufmann, Burlington (2016)
124. Wong, T.T.: A hybrid discretization method for naïve Bayesian classifiers. Pattern Recogn. **45**(6), 2321–2325 (2012)
125. Xue, B., Zhang, M., Browne, W.N.: Particle swarm optimization for feature selection in classification: A multi-objective approach. IEEE Trans. Cybern. **43**(6), 1656–1671 (2013). https://doi.org/10.1109/tsmcb.2012.2227469
126. Xue, B., Zhang, M., Browne, W.N., Yao, X.: A survey on evolutionary computation approaches to feature selection. IEEE Trans. Evol. Comput. **20**(4), 606–626 (2016). https://doi.org/10.1109/tevc.2015.2504420
127. Zhao, H.: A multi-objective genetic programming approach to developing Pareto optimal decision trees. Decis. Support. Syst. **43**(3), 809–826 (2007). https://doi.org/10.1016/j.dss.2006.12.011
128. Zhao, J., Fernandes, V.B., Jiao, L., Yevseyeva, I., Maulana, A., Li, R., Back, T., Tang, K., Emmerich, M.T.: Multiobjective optimization of classifiers by means of 3D convex-hull-based evolutionary algorithms. Inf. Sci. **367-368**, 80–104 (2016). https://doi.org/10.1016/j.ins.2016.05.026
129. Zheng, K., Wang, X.: Feature selection method with joint maximal information entropy between features and class. Pattern Recogn. **77**, 20–29 (2018)
130. Zhou, A., Qu, B.Y., Li, H., Zhao, S.Z., Suganthan, P.N., Zhang, Q.: Multiobjective evolutionary algorithms: a survey of the state of the art. Swarm Evol. Comput. **1**(1), 32–49 (2011). https://doi.org/10.1016/j.swevo.2011.03.001
131. Zitzler, E., Thiele, L.: Multiobjective evolutionary algorithms: a comparative case study and the strength Pareto approach. IEEE Trans. Evol. Comput. **3**(4), 257–271 (1999) https://doi.org/10.1109/4235.797969
132. Zitzler, E., Deb, K., Thiele, L.: Comparison of multiobjective evolutionary algorithms: empirical results. Evol. Comput. **8**(2), 173–195 (2000)

No Free Lunch Theorem: A Review

**Stavros P. Adam, Stamatios-Aggelos N. Alexandropoulos,
Panos M. Pardalos ⓘ, and Michael N. Vrahatis**

Abstract The "No Free Lunch" theorem states that, averaged over all optimization problems, without re-sampling, all optimization algorithms perform equally well. Optimization, search, and supervised learning are the areas that have benefited more from this important theoretical concept. Formulation of the initial No Free Lunch theorem, very soon, gave rise to a number of research works which resulted in a suite of theorems that define an entire research field with significant results in other scientific areas where successfully exploring a search space is an essential and critical task. The objective of this paper is to go through the main research efforts that contributed to this research field, reveal the main issues, and disclose those points that are helpful in understanding the hypotheses, the restrictions, or even the inability of applying No Free Lunch theorems.

1 Introduction

Optimization problems occurring in various fields of science, computing, and engineering depend on the number of parameters, the size of the solution space and, mainly, on the objective function whose definition is critical as it largely determines

---

S. P. Adam
Department of Informatics and Telecommunications, University of Ioannina, Arta, Greece

Computational Intelligence Laboratory – CILab, Department of Mathematics, University of Patras, Patras, Greece
e-mail: adamsp@upatras.gr

S.-A. N. Alexandropoulos · M. N. Vrahatis (✉)
Computational Intelligence Laboratory – CILab, Department of Mathematics, University of Patras, Patras, Greece
e-mail: alekst@math.upatras.gr; vrahatis@math.upatras.gr

P. M. Pardalos
Department of Industrial & Systems Engineering, University of Florida, Gainesville, FL, USA
e-mail: pardalos@ufl.edu

© Springer Nature Switzerland AG 2019
I. C. Demetriou, P. M. Pardalos (eds.), *Approximation and Optimization*,
Springer Optimization and Its Applications 145,
https://doi.org/10.1007/978-3-030-12767-1_5

57

the level of difficulty of the problem. Hence, defining and solving an optimization problem is sometimes an extremely difficult and demanding task. Researchers from various fields have been involved in solving optimization problems either as this constitutes part of their main research or because the problem they face can be tackled by an optimization one. The research efforts on this matter have permitted the elaboration of numerous methods and techniques, built on solid mathematical concepts, whose application produced significantly good results.

However, contrary to any opposite claim, none of these methods has proven to be successful to all types of the problems it was applied. This argument has been the objective of important theoretical work carried out by David Wolpert which gave rise to the well-known *No Free Lunch* (NFL) theorem. Briefly, the NFL theorem states that: *"averaged over all optimization problems, without re-sampling all optimization algorithms perform equally well."* Besides optimization, the NFL theorem has been successfully used to tackle important theoretical issues pertaining supervised learning in machine learning systems. Actually, the NFL theorem has become a suite of theorems which has given significant results in various scientific fields where searching for some optimal solution is an important issue.

The NFL theorems constitute an important theoretic development which marked the limits of the range of successful application for a number of search, optimization, and supervised learning algorithms. At the same time the formulation of these theorems has provoked controversial discussions [4, 36, 44, 45] regarding the possibility to invent and effectively use general purpose algorithms in various fields where only a limited view of the real-world problem exists.

In this paper we aim at presenting a review on the most sound research work published by several researchers on this matter including its impact on the most important fields, that is, optimization and supervised learning. Other existing fields of interest such as user interface design [24], network calculus [8] are worth of merit but they are out of the scope of this review. The emphasis of this review will be, mainly, on the critical questions which promoted the development of NFL theorems as well as on the issues that proved to be important: namely for (a) *optimization*, (b) *searching*, and (c) *supervised learning*.

The rest of this paper is structured as follows. Section 2 provides a review of the early concepts and constructs that underpinned the definition of the NFL theorems. Section 3 covers the main research efforts of Wolpert establishing NFL for optimization and search. In Section 4 we survey the more recent work of Wolpert which clarifies older concepts while offering some new results on this field. Next, Section 5 is dedicated to the main research carried out by several researchers on NFL for optimization and evolutionary algorithms. Part of the research surveyed concerns the cases where NFL theorems do not apply and researchers have proved the existence of *"Free Lunches."* In Section 6 we describe the main research efforts on NFL theorems for supervised learning. The paper ends in Section 7 with a synopsis and some concluding remarks.

2 Early Developments

As noted by David Wolpert [56], the first attempt to underline the limits of inductive inference was made by the Scottish philosopher David Hume in 1740 in his seminal work *"A treatise of human nature"* [26, 27]. Hume wrote that:

> Even after the observation of the frequent conjunction of objects, we have no reason to draw any inference concerning any object beyond those of which we have had experience.

In the machine learning context this can be stated as follows:

> It is not reasonable to believe that the generalization error of a classifier-generalizer on test data drawn off the training set correlates with its performance on the training set itself by simply considering a priori information on the real world.

Wolpert based his theoretical work on earlier developments elaborated in his paper *"On the connection between in-sample testing and generalization error"* [55]. In this paper the generalization error is taken as the *off-training set* (OTS) error and the question addressed concerns its correlation with the error produced using in-sample testing. Moreover, Wolpert tackles the question of how *"...to take into account the probability distribution of target functions in the real world"* as any theory of generalization is irrelevant concerning its applicability on real- world problems if it does not tackle the previous problem. Some, but not all, of the important issues arising in this paper are:

(a) *"Can one prove inductive inference from first principles?"* In other words, given the performance of a learning algorithm on the training data set is it possible to obtain information on its ability to provide an exact representation of the target function for examples outside the data set?
(b) If one cannot answer the previous question then, what are the assumptions on the distribution of real-world data (the target function) can help with the generalization for training algorithms, such as back-propagation, which aim to minimize the error on the training data?
(c) Is there a mathematical basis of estimating when over-training occurs and proceed in modifying the learning algorithm in order to bound the effects of such over-training?
(d) Is it possible to express in mathematical terms the ability of a training set to faithfully represent the distribution over the entire data space?
(e) What are the hypotheses under which non-parametric statistics techniques, such as cross-validations, which are designed to choose between learning algorithms, succeed to diminish the generalization error?

In addressing these matters, the formalism proposed seems to extend the classical Bayesian formalism using the hypothesis function, i.e., the distribution of the data set as learned by the generalizer. The mathematical formalism adopted proposes a way to match the degree to which the distribution derived by the learning algorithm matches the distribution of the training data and it can be used to tackle various generalization issues such as over-training and minimum number of parameters for the model. From another point of view this formalism is proposed with the aim to express in mathematical terms the assumptions made by a generalizer so that the used model best fits the training set representing the real world. As a result the elaboration of important theoretical proofs proposes a solid basis for tackling several issues in machine learning and gives rise to the development of concepts such as the NFL theorems.

The first and foremost contributions of Wolpert concerning NFL theorems were presented in the papers [56, 57]. In this set of two papers, namely:

(i) *"The lack of a priori distinctions between learning algorithms"* and
(ii) *"The existence of a priori distinctions between learning algorithms,"*

Wolpert develops his theory and formulates the NFL theorems. In the former, he discusses the hypothesis that given any two learning algorithms one cannot claim having any prior information that these algorithms are distinct as far as the performance of these algorithms on specific class of problems is concerned. In the latter paper, Wolpert unfolds the arguments concerning the inverse assumption, i.e., there are prior distinctions regarding the performance of any two algorithms. These two papers deal with supervised learning but the theoretical constructs were applied to multiple domains where two different algorithms compete as for which performs better for a class of problems and associated error functions.

Focusing on supervised learning, in the first of the previously mentioned papers the concept of *"off-training set"* (OTS) is defined and the associated performance measure of the supervised learning algorithm is proposed. The mathematical formalism used is based on the so-called *extended Bayesian formalism* and is refined in order to take into account the generalization error, the cost function, and their relation to the learning algorithm while providing the necessary hypotheses for the training sets and the targets. In the sequel the probability of some cost "c" of the learning algorithm associated with the loss function is proposed as follows:

$$P\,(c|d) = \int df\,dh\,P\,(h|d)\,\,P\,(f|d)\,\,M_{c,d}\,(f,h)\,,$$

which is considered to be the inner product between the infinite dimensional vectors $P(f|d)$ and $P(h|d)$ representing the target and the hypothesis functions, respectively. This inner product quantity is maximized if the target function f and the hypothesis function h given the training data d are close enough to each other, i.e., they are aligned. Given two learning algorithms (generalizers) A and B, an important question to be answered deals with the comparison of these two algorithms in terms of how the set F_1 of target functions f for which A beats B compares with the corresponding set F_2 of the target functions f for which algorithm B outperforms A. As it is stated in [56]: *"in order to analyze this issue it is proposed to compare the average over f of f-conditioned probability distributions for algorithm A to the same average for algorithm B. Then the relationship between these two averages is used to compare the sets F_1 and F_2 ."*

In the second paper: *"The existence of a priori distinctions between learning algorithms"* [57] Wolpert, besides revisiting the theorems and some examples of the first paper, examines the NFL theorems with respect to cross-validation and the so-called *head-to-head minimax* behavior that is the case where for an algorithm A there exist comparatively few target functions for which A is slightly worse than algorithm B and comparatively few target functions in which algorithm A is superior to algorithm B. Moreover, he develops an extension of his theory by considering averaging over generalizers rather than targets. This means that instead of characterizing two algorithms by averaging over targets, namely f, ϕ, $P(f)$, or $P(\phi)$, holding the hypothesis, $P(h|d)$, fixed it is tentative to consider alternative results where one holds one of the entities concerning the targets, fix and average over the hypothesis entities. For this case, Wolpert formulates some additional theorems and finally he examines the case when the loss function $L(\cdot|\cdot)$ is non-homogenous and thus the NFL theorems do not apply as one can make a priori distinctions between algorithms.

As a conclusion it is stated in [57] that these two papers investigate some of the behavior of OTS error. In particular, they formalize and investigate the concept that *"if you make no assumptions concerning the target, then you have no assurances about how well you generalize."*

3   No Free Lunch for Optimization and Search

Another direction of research for applying the ideas of the NFL theorems, as presented above, concerns the domain of optimization. The work *"No free lunch theorems for optimization"* [62] published by Wolpert and McReedy deals with this matter based on two technical reports produced by the authors at the Santa Fe Institute. The first technical report published in [35] with the title *"What makes an optimization problem hard?"* raises the question: *"Are some classes of combinatorial optimization problems intrinsically harder than others, without regard to the algorithm one uses, or can difficulty be assessed only relative to a particular algorithm?"* The second technical report [61], entitled: *"No free lunch theorems for search"* focuses on proving that all algorithms searching for an optimum of an optimization problem, i.e., an extremum of an objective function, performs exactly the same, no matter the performance measure used, when taking the average over all possible objective functions.

The work of Wolpert and McReedy *"No free lunch theorems for optimization"* [62], sets up a formalism for investigating the relation of the effectiveness of optimization algorithms and the problems they are solving. The NFL theorems developed in the paper establish that the successful performance of any optimization algorithm on one class of problems is counterbalanced by its degraded performance on another class of problems. A geometric interpretation is provided concerning the meaning of the fitness of an algorithm to cope with some optimization problem. Moreover, as mentioned in the previous technical reports the authors examine applications of NFL theorems to information-theoretic aspects of optimization as well as to defining measures of performance for optimization benchmarks.

Given the multitude of black-box optimization techniques available, the authors try to provide the formalism for tackling the following problem: *"is there a relationship between how well an algorithm performs and the optimization problem on which it is run?"* This problem can be cast in several other such as:

(a) What are the mathematical constituents of optimization theory one needs to know before deciding on the necessary probability distributions to be applied?
(b) Are information theory and Bayesian analysis suitable for understanding the previous issues?
(c) Given the performance results of a certain algorithm on a certain class of problems can one provide a priori generalization of these results on other classes of problems?
(d) Is there a suitable measure of such generalization? Can one evaluate the performance of algorithms on problems so that he is able to compare those algorithms?

The formalism developed by the authors is articulated around the following concepts:

(i) A sample of size m is a set of m distinct points visited by the algorithm and is denoted by

$$d_m = \left\{ \left(d_m^x(1),\, d_m^y(1)\right),\, \left(d_m^x(2),\, d_m^y(2)\right),\, \ldots,\, \left(d_m^x(m),\, d_m^y(m)\right) \right\},$$

where $d_m^x(i)$ denotes the $\mathscr{X}$ value of the i th element of the sample and $d_m^y(m)$ is the associated cost, i.e., the $\mathscr{Y}$ value.

(ii) An optimization algorithm α is a mapping from previously visited sets of points to a single new point in $\mathscr{X}$, i.e.,

$$\alpha:\, d \in \mathscr{D} \to \left\{ x \mid x \notin d^x \right\},$$

where $\mathscr{D}$ denotes the space of all (m-sized) samples and α is deterministic in the sense that every sample maps to a unique new point.

(iii) The performance of an algorithm after m iterations is a function $\Phi(d_m^y)$ of the sample.

(iv) Given the space of all cost functions, i.e., optimization problems $\mathscr{F}$ the distribution:

$$P(f) = P\left(f(x_1),\, f(x_2),\, \ldots,\, f(x_{|\mathscr{X}|})\right),$$

defined over $\mathscr{F}$ gives the probability that each $f \in \mathscr{F}$ is the actual optimization problem at hand.

(v) The performance of an optimization algorithm α on a cost function f after m iterations is measured with $P(d_m^y \mid f, m, \alpha)$.

Let us consider the problem:

> Suppose that $F_1 \subseteq \mathscr{F}$ is the set of problems for which an algorithm α_1 performs better than algorithm α_2 and $F_2 \subseteq \mathscr{F}$ denotes the set for which α_2 performs better than α_1. How can one compare these two sets?

The answer provided by the authors relies on the sum of $P(d_m^y \mid f, m, \alpha_1)$ and the sum of $P(d_m^y \mid f, m, \alpha_2)$ over all f, i.e., over all problems. The following theorem as formulated in this paper addresses the previous problem.

Theorem 1 *For any pair of algorithms α_1 and α_2,*

$$\sum_f P\left(d_m^y \mid f, m, \alpha_1\right) = \sum_f P\left(d_m^y \mid f, m, \alpha_2\right).$$

In the theorem the problem is considered to be fixed over time. If the cost function is time-dependent in the sense that, while the problem is initially expressed with some cost function f_1 which is present when sampling the first value in $\mathscr{X}$, then this function is deformed before any subsequent iteration of the optimization algorithm. If deformation is represented with the mapping $T : \mathscr{F} \times \mathscr{N} \to \mathscr{F}$, and $T = T_i$, then $f_{i+1} = T_i(f)$ and the following theorem can be formulated:

Theorem 2 *For all d_m^y, D_m^y, $m > 1$, algorithms α_1 and α_2, and initial cost functions f_1*

$$\sum_T P(d_m^y \mid f_1, T, m, \alpha_1) = \sum_T P(d_m^y \mid f_1, T, m, \alpha_2),$$

and

$$\sum_T P(D_m^y \mid f_1, T, m, \alpha_1) = \sum_T P(D_m^y \mid f_1, T, m, \alpha_2).$$

One of the implications of the NFL theorems discussed by the authors deals with the geometric perspective of NFL. In this perspective consider the space $\mathscr{F}$ of all cost functions and the probability of obtaining a certain d_m^y defined by the relation:

$$P(d_m^y \mid m, \alpha) = \sum_f P(d_m^y \mid m, \alpha, f) \, P(f),$$

with $P(f)$ being the prior probability that the optimization problem at hand has cost function f. As noted by the authors the previous sum can be considered as an inner product in $\mathscr{F}$. Hence, if we define the vectors $\vec{v}_{d_m^y, \alpha, m}$ and $\vec{p}$ by their f components, respectively:

$$\vec{v}_{d_m^y, \alpha, m}(f) \equiv P(d_m^y \mid m, \alpha, f), \quad \text{and} \quad \vec{p} \equiv P(f),$$

then it holds that:

$$P(d_m^y \mid m, \alpha) = \overrightarrow{v}_{d_m^y, \alpha, m} \cdot \overrightarrow{p} .$$

The authors note that this equation provides a geometric interpretation of the optimization process. Hence, d_m^y represents the desired sample and m is taken as a measure of the computational effort needed for the algorithm. Moreover, if the vector $\overrightarrow{p}$ represents the prior which includes all knowledge about the cost functions, then the last equation formulates in mathematical terms that: "*the performance of an algorithm is determined by the magnitude of its projection on $\overrightarrow{p}$ or in other words by how aligned $\overrightarrow{v}_{d_m^y, \alpha, m}$ is with the problem's vector $\overrightarrow{p}$.*" With respect to the geometric view the NFL result that $\sum_f P(d_m^y \mid f, m, \alpha)$ is independent of α means that for any particular d_m^y and m, all algorithms have the same projection onto the uniform $P(f)$ represented by the diagonal vector $\overrightarrow{1}$.

Moreover, the authors investigate the relationship of the above results with information-theoretic aspects of optimization and provide measures of performance for assessing the efficacy of a certain optimization algorithm. Finally, minimax distinctions between search algorithms are discussed and some performance measures for search algorithms are provided.

4   More Recent Work of Wolpert

The work of Köppen, Wolpert, and McReedy *"Remarks on a recent paper on the No Free Lunch Theorems"* [33] is a letter reconsidering a previous work of Köppen [32] with the title *"Some technical remarks on the proof of the No Free Lunch theorem."* In this letter the authors, following suggestions made in [32], provide a short proof of the NFL theorems while correcting a wrong claim made in [32] about circular reasoning of the original proof of the NFL theorems in [61, 62].

Hereafter, let us give some details on this theorem, as presented in [61, 62]; its proof is important for many papers on NFL theorems. First, consider two finite sets X and Y together with the set of all cost functions $f : X \rightarrow Y$. Moreover, for a positive integer m such that $m < |X|$ let $d_m = \{(d_m^x(i), d_m^y(i) = f(d_m^x(i)))\}$ i.e., the points sampled by the algorithm in m steps, with $i = 1, 2, \ldots, m$ $d_m^x(i) \in X$ $\forall x$ and for any i, j it holds that $d_m^x(i) \neq d_m^x(j)$. Let a denote the search algorithm of interest, which is a deterministic "blind" algorithm assigning to every possible d_m an element of X which is not already in the d_m^x. This means that,

$$d_{m+1}^x(m + 1) = a[d_m] \notin \{d_m^x\}.$$

Let $Y(f, a, m)$ denote the sequence of the m values of Y produced by the algorithm a to f after m successive steps and $\delta(\cdot, \cdot)$ is the Kronecker delta function giving 1 if its arguments are identical and 0 otherwise. Then the following lemma holds:

Lemma 1 *For any algorithm a and any d_m^y,*

$$\sum_f \delta\left(d_m^y,\ Y(f, m, a)\right) = |Y|^{|X|-m}.$$

Thus, if $c(\cdot)$ is some performance measure assigning a real value to any set d_m^y and $k \in \mathbb{R}$ is a performance value, then the theorem in question is:

Theorem 3 *For any two deterministic algorithms a and b, any value $k \in \mathbb{R}$, and any performance measure $c(\cdot)$,*

$$\sum_f \delta\left(k,\ c(Y(f, m, a))\right) = \sum_f \delta\left(k,\ c(Y(f, m, b))\right).$$

Besides considering the proof of this theorem, in this letter, the authors take the chance to defend NFL theorems against what they call a rather nihilistic view that algorithms of universal applicability would not exist. NFL theorems should be considered as a research topic and not as simply some convenient or inconvenient result. Hence, they propose that a more open minded view should prevail in order to investigate the limits of NFL theorems as well as the potential issues arising by their application in various domains.

It is worth noting, here, a relatively more recent work of Wolpert [60] entitled: *"What the No Free Lunch Theorems Really Mean? How to Improve Search Algorithms?"* In this research report the author reconsiders the main ideas of his work on NFL as far as search algorithms are concerned. Wolpert insists on analyzing the issue that while the NFL theorems have strong implications whenever a uniform distribution of the cost function over the optimization problems is adopted, this is not meant to support the use of such a distribution when one has to solve an optimization problem. Trying to clarify what the NFL really mean in order to improve search algorithms, Wolpert analyzes some kind of *"deep formal relationship between supervised learning and searching."* As a result of the analysis of this relationship there are NFL theorems for both search and supervised learning and so there are various ways of reusing techniques first developed in supervised learning for guiding search. A number of experiments are presented which confirm the effectiveness of search algorithms built upon these concepts.

5 NFL for Optimization and Evolutionary Algorithms

5.1 No Free Lunches and Evolutionary Algorithms

The NFL theorems have attracted the interest of the scientific community and keep this interest unchanged. On the other hand, there has been occasionally the bone of contention between some researchers. Such conflicting positions are listed in

Perakh's essay [42]. In particular, one may note the position of Orr [37] regarding the NFL theorems, which was presented on the occasion of the publication of William Dembski's book [10]. Orr stated that:

> ...NFL theorems compare the effectiveness of evolutionary algorithms and look at how often such an algorithm can detect the target, within a certain number of steps...

Orr underlined some very useful observations regarding NFL theorems in relation with Darwinian theory and this has been the essence of the difference between Orr and Dembski. More precisely, Orr claims that *evolution* according to Darwin's theory cannot be seen as a search process and therefore, contrary to Dembski, one cannot claim that Darwinism constitutes a search algorithm. It is evident that NFL theorems do not exclude evolutionary process defined according to the Darwinian theory. Hence, evolutionary algorithms can be appropriate used for search and they are capable to overcome a random search algorithm.

In [42] Perakh gave a popularized interpretation of NFL theorems. This interpretation is presented hereafter along with some useful comments and remarks as given by the author. Suppose that A and B are two search algorithms, exploring the same search space. The algorithms explore the search space by moving from one point to another, selecting points either randomly or following a specific order. Each algorithm performs a certain number of moves. At any point visited the algorithm computes the value of the fitness function and so after, say, k steps the algorithm provides k measurements, which constitute what is called a sample.

In essence, this sample is nothing more than a table in which the values of the fitness function are recorded for each search point. However, an important question arises: *"could two algorithms return the same sample, given that they have selected the same number of points?"* Obviously, the samples computed by two arbitrarily chosen algorithms are not expected to be the same. This argument can be easily understood if one considers it in terms of probabilities.

Specifically, Perakh notes that:

> The probability of a sample (i.e., a table), of size k, produced by an algorithm, say A, differs from the probability that the same sample is produced by another algorithm B, for the same number of steps of the algorithms.

However, the first NFL theorem states that if the search results of the two algorithms are not compared for a particular fitness space but averaged over all possible search spaces, then the above probabilities of obtaining the same sample are equal for any pair of algorithms.

It is worth to underline that the NFL theorems are valid regardless how many times the algorithms are used to complete a search of the underlying problem space or which fitness function values are returned by the different search points. Another point that is worth paying attention is that NFL theorems make no claim about the relative performance of the algorithms, as defined in [42], for a particular search space. As a result, in terms of performance, any algorithm could be much better than any other "competitor."

Despite the fact that NFL theorems are valid for evolutionary algorithms, in [58] it is argued that this may not stand for the case of co-evolutionary algorithms and so *"Free Lunches"* are possible. The NFL framework for the case of co-evolutionary algorithms as described in [49] is given next.

The statement relative to *"the average performance of the algorithms,"* mentioned in the previous paragraphs and references therein, is meaningless without the definition of how this performance is measured. In other words, a very important issue is to define the metrics that one should use in order to effectively measure and compare the performance of the algorithms.

In addition, some other important questions may arise, such as:

(a) *Are there any classes of co-evolution for which there exist NFL theorems?*
(b) *For which co-evolution classes there can be Free Lunches?*

According to the literature [63] these questions are difficult to be answered and they still remain open problems.

Some recent research efforts regarding NFL theorems and black-box optimization have shown that there are co-evolutionary problems with No Free Lunches while Free Lunches are present in the context of other co-evolutionary problems. More precisely, in their work [49], Service and Tauritz present a NFL framework for classes of co-evolutionary algorithms. What is important in this work is the classification of co-evolutionary algorithms based on the solutions they seek. In the co-evolutionary algorithms framework defined in this work, the type of the solutions, or the corresponding individuals, that are effectively considered as solutions to the problem, depend, exclusively, on the type of the problem for which the co-evolutionary algorithm is designed. Note that the different solution concepts are related to the cooperative co-evolution case, the Nash equilibrium case, the maxmin case, etc.

The authors define the so-called weak preference relation which is a relatively simple way of measuring the performance of co-evolutionary algorithms and so it constitutes a metric. This metric is different than the one originally defined by Wolpert and Macready in their work *"Co-evolutionary Free Lunches"* [63].

The framework developed by Service and Tauritz can be considered as a combination of concepts and definitions originating from two theoretical frameworks. The first framework deals with the original NFL theorems [62] for search algorithms and the other for concerns co-evolutionary algorithms [18]. This fusion is done with respect to the consistency of both frameworks. Moreover, in [49] the authors showed that in co-evolution there are Free Lunches. In consequence, the important question that remains to be answered is: *"in which classes of co-evolutionary algorithms there are Free Lunches?"* and further studies are needed to explore additional classes of co-evolution.

5.2   No Free Lunches and Meta-Heuristic Techniques

It is well known that particle swarm optimization methods [41] have greatly contributed to the field of mathematical optimization. These swarm- based methods consist of a number of individuals who guide the optimization process through their collective behavior in order to attain an optimal solution. A great advantage of these methods is that, under suitable conditions and assumptions, they are capable to avoid local minima and ensure convergence of the algorithm to some globally optimal solution. However, convergence analysis of swarm optimization algorithms still remains an active research areas. The most important schemes that have been define and used include: *"Particle Swarm Optimization"* (PSO) [14], *"Ant Colony Optimization"* (ACO) [11], *"Firefly Algorithm"* (FA) [64], *"Artificial Bee Colony algorithm"* (ABC) [28], *"Bat Algorithm"* (BA) [65], *"Cuckoo Search"* (CS) [67], among others.

These methods, also called meta-heuristic techniques, involve exploration and exploitation; two specific search processes which under appropriate conditions "control" the swarm in order to avoid local minima of the fitness function. The applications of the above swarm-based schemes are many and belong to different scientific fields. More details on these can be found in [19], especially concerning engineering and industrial applications. In recent years, application of meta-heuristic techniques has constantly increased and has entered the field of art [1, 2, 12, 47, 52]. More specifically, meta-heuristic techniques have been applied in the tasks of Crowd simulation, Human swarming, and Swarmic art.

Meta-heuristic techniques or meta-models, such as those proposed in [38–41, 43, 50], are used in many cases of evolutionary computing techniques [15–17, 20, 34], in order to create faster optimization algorithms. Especially, in cases where data sets are incomplete or imbalanced or the objective function is computational costly, the meta-heuristic procedures provide alternative, effective, and efficient solution to the optimization problem. Specifically, these techniques are high-level heuristic processes that aim at choosing or creating meta-heuristic search models to resolve more efficiently optimization problems. As noted in [4], under some mild conditions with respect to objective functions, the surrogate algorithms achieve global convergence [5]. In addition, these meta-models are not plagued by damaging features of classic optimization methods, such as the calculation of derivatives. As a consequence, meta-models outperform classic methods, enabling them to be effectively and efficiently deployed in a variety of applications, such as [25, 30, 53].

An important discussion concerning NFL for meta-heuristics is proposed by Yang [66]. In this work the author notes that NFL theorems deal with the average performance of optimization algorithms on all existing problems. Nevertheless, in many real problems this does not hold, as the theoretical requirements are strict and they cannot be applied, in practice. As a consequence, this situation results in getting Free Lunches, and what needs to be determined is the performance of specific algorithms in particular classes of problems. Hence, in such cases there may exist algorithms that are significantly better than others for a particular class of problems. This phenomenon, i.e., the non-validity of NFL theorems, often occurs when applying meta-heuristic approaches, as the primary NFL theorems concern algorithms searching for individual solutions while population-based meta-heuristic approaches explore simultaneously different parts of the search space and, in this sense, they are considered dealing with sets of solutions. As an example one may consider the cases of genetic algorithms or PSO. A similar situation is encountered in multi-objective optimization, where some algorithms are found to outperform others on specific problems, thus, giving rise to Free Lunches [9].

The theoretical results of NFL theorems while being very important for mathematical optimization with significant theoretical impact, however, incite a number of questions related with practical applications such as: *"What is the position and the opinion of optimization algorithm designers on the practical validity and the applicability of NFL theorems?"*

Yang [66] provides an answer to this question and classifies developers of optimization algorithms in three groups:

(a) A large part of researchers believe that the conditions set by NFL theorems cannot be applied in practice and therefore they do not accept them.
(b) Researchers in the second category accept the validity of NFL theorems but they believe that for specific types of problems there exist optimal algorithms. So, they focus on finding such algorithms for particular classes of problems.
(c) The last group claims that NFL theorems do not hold for continuous problems or for problems belonging to the NP-hard class. Therefore, they focus on discovering problems for which NFL theorems do not apply and hence on defining Free Lunches.

The appeal and the controversies caused by NFL theorems led a large part of the scientific community to re-examine these theorems and restate them in several equivalent forms. The studies resulting from this trend have led to the creation of many frameworks for black-box search algorithms such as the framework proposed by Schumacher et al. [48]. These authors studied the length of the problem description and they concluded that the NFL results as initially formulated by Wolpert are valid not only for the set of all functions but even for smaller sets. Hence, NFL results are independent of whether the set of functions is compressible or not. Finally, the authors conclude that the results of NFL theorems are best maintained in the case of the permutation closure of a single function.

The variety of scientific fields where NFL theorems have been applied made more and more researchers study and apply these theorems which led to the proposition of various extensions of NFL theorems. It is worth mentioning that Auger and Teytaud [4] proposed extensions of NFL theorems related to infinite spaces both countable and uncountable. In addition, they studied the design of optimal heuristic optimization models. According to the original work of Wolpert and Macready [62], the NFL theorems for optimization concern finite search spaces. So, in order to extend the theorems to infinite search spaces, stochastic terms and procedures are introduced. The authors demonstrated that in the case of infinite countable spaces, the physical extension of the NFL theorems does not hold. In addition, for their proof they defined some distributions of the fitness functions, which lead to equal performance for all heuristic search techniques.

The above proof resulted in Free Lunch theorems based on a random fitness function and involves random search spaces. An additional contribution made in [4] deals with designing optimal algorithms for random fitness functions regarding a black-box optimization framework. In particular, the authors presented an optimal algorithm based on the Bellman's decomposition principle [6], for a certain number of algorithm iterations and a given distribution of fitness. Moreover, for the design procedure and the experiments conducted, the *"Monte-Carlo planning"* algorithm [31] and the *"Upper Confidence Tree"* algorithm [51] were used. Following these research results one may, reasonably, put forward the question: *"Is the improvement proposed by Auger and Teytaud just of theoretical importance or it can be applied in practical situations in acceptable computational time?"*

Hereafter, in order to present some of the results of Auger and Teytaud [4], more formally, we recall the necessary notation adopted in [4]. Let $\mathscr{X}$ denote the search space and $\mathscr{Y}$ its codomain for a given objective function f. For any integer $m \in \{1, 2, \ldots, |\mathscr{X}|\}$ let $(x_1, x_2, \ldots, x_m)$ be the vector of the first m iterates of a search algorithm and let $(f(x_1), f(x_2), \ldots, f(x_m))$ be the vector of the associated objective values. The performance of an algorithm a after m iterations is given by measuring the vector of cost values denoted by $Y(f, m, a) = \langle f(x_1), f(x_2), \ldots, f(x_m) \rangle$.

Using the previous notation, NFL theorems imply the following results for $\mathscr{X}$ any finite domain, $\mathscr{Y}$ its codomain, two search algorithms a and b, any number of iterations m and, finally, any objective function f and p any random permutation uniformly distributed (among all permutations) over $\mathscr{X}$: the random vectors:

$$Y(f \circ p, m, a) = \langle f \circ p(x_1), \ f \circ p(x_2), \ \ldots, \ f \circ p(x_m) \rangle,$$

and

$$Y(f \circ p, m, b) = \langle f \circ p(x_1), \ f \circ p(x_2), \ \ldots, \ f \circ p(x_m) \rangle,$$

follow the same distribution.

Moreover, let $\mathscr{X}$ be a countably infinite space and without loss of generality let $\mathscr{X} = \mathbb{N}$. If one is able to provide a non-trivial measurable objective function f, then the following proposition holds:

Proposition 1 *Assume that $\mathscr{NFL}(\mathbb{N}, p, f)$ is a No Free Lunch, and*

$$f(i) = (-1)^{i+1} i, \quad \forall i \in \mathbb{N}.$$

Then there is no random permutation p such that $\mathscr{NFL}(\mathbb{N}, p, f)$ holds. Consequently, the $\mathscr{NFL}(\mathbb{N}, f)$ does not hold.

Table 1 Number of citations of the references related to the NFL theorems for Optimization and Evolutionary Algorithms issues presented in Section 5

Contribution	Total citations	Citations per year
Schumacher et al. [48]	190	11.18
Droste et al. [13]	150	9.38
Perakh [42]	6	0.40
Griffiths and Orponen [23]	9	0.69
Service and Tauritz [49]	4	0.40
Poli and Graff [44]	35	3.18
Auger and Teytand [4]	70	8.75
Yang [66]	34	5.67

Among different theoretical results, one may stick to the following Continuous Free Lunch theorem which is considered to be the main result of Auger and Teytaud in [4].

Theorem 4 (Continuous Free Lunch) *Assume that f is a random fitness function with values in $\mathbb{R}^{[0,1]}$. Then $\mathscr{GNFL}([0, 1], f)$ does not hold.*

In the above theorem $\mathscr{X}$ is considered to be a continuous domain and without loss of generality $\mathscr{X} = [0, 1]$ and $\mathscr{Y} = \mathbb{R}$. Moreover, the notation $\mathscr{GNFL}$ is used for a weaker NFL which does not restrict the fitness function to the compositional form $f \circ p$.

Remark 1 Let f be a random fitness. Then $\mathscr{GNFL}(\mathscr{X}, f)$ holds if and only if for any $m \in \mathbb{N}$ (smaller than $|\mathscr{X}|$ when $\mathscr{X}$ is finite) and any two optimization algorithms a and b, $Y(f, m, a)$ and $Y(f, m, b)$ follow the same distribution.

In Table 1, we provide information about the number of citations[1] received by the most significant contributions concerning the field of the NFL theorems for Optimization and Evolutionary Algorithms. This citation analysis can be considered as an additional information about the importance and contribution of these works in the field of the NFL theorems.

Designing an optimization algorithm that will be more effective than other optimization schemes is a very difficult process and requires a number of conditions. *"Multidisciplinary Design Optimization"* is a problem that is based on the best architecture selection. In such a context, it can be easily understood that obtaining the most efficient design scheme requires testing and may lead to errors. However, the trial-and-error procedure is not appropriate as it is a costly computational process. Vanaret et al. [54] proposed a general design process that avoids the above problem, as well as the inherent complexity that exists in such applications.

---

[1] Source: Google Scholar.

In most cases of Multidisciplinary Design Optimization, having efficient optimization algorithms is seriously restricted by the complexity of the objective function which is due to the fact that several different architectures are used for the design task. So, it is of primary importance to dispose a methodology that can be applied in all design cases and alleviate this disadvantage. In [54] this is accomplished through a replacement function that can be calculated much more easily than the original one. The authors propose a scalable replacement model through which the architectures can be evaluated easier and thus the choice made more appropriate. Through their experimental results it is clear that the performance of an architecture model depends significantly on the dimension of the original problem. Therefore, as stated by the NFL theorems, there is no architecture that is significantly more efficient than all the others, when dealing with problems of the same dimension. The authors adopt the *"Multidisciplinary Feasible"* and the *"Individual Disciplinary Feasible"* architecture models as the more representative among different architecture models for Multidisciplinary Design Optimization problems. Nevertheless, more architecture models need to be explored in the future.

Kimbrough et al. [29] studied several cases of optimization with constraints using population-based optimization algorithms. In particular, in their research they used genetic algorithms regarding two populations, those of feasible solutions and those of non-feasible ones. Theoretically, in a simple, typical scheme of genetic evolution, the individuals evaluated as feasible solutions would be the only ones that would take part in the evolution of the population and so, in the final formulation of the solution. However, this theoretical provision is not valid in [29] as in this case Kimbrough et al. use feasible solutions in improving the values of the objective function, while non-feasible solutions are used to correct the penalties caused by their constrains.

In order to ensure the smoothness of the optimization process, namely the evolution of the populations, the authors defined a metric distance between the two populations, both among the individuals and the populations' centroids. An important detail that has to be underlined is that the centroids of the two populations are approaching each other during the evolution.

At first sight, it might seem strange to maintain a whole population of infeasible solutions. However, a closer look reveals the usefulness of this position as this population is free to move to space areas, where the feasible solutions cannot, and thus to explore the limited search areas. The authors studied specific problems and spaces and showed that the conclusions of the NFL theorems regarding the equivalence of the black-box search algorithms do not hold. Furthermore, they shown that the NFL theorems do not hold for problems with constraints and specifically in many practical problems where the restrictions are fixed.

The evolutionary computing scheme adopted by Kimbrough et al. [29] is an elegant mechanism which permits to show that there exist constraint optimization problems for which NFL results do not hold. The interested reader is invited to refer to the work [29].

As a supplement of the above research works along with both the theoretical arguments and the conclusions of the specific problem classes mentioned above, Droste et al. in [13] provide some realistic remarks based on the computational complexity of heuristic optimization algorithms. The authors claim that NFL theorems are not possible in the case of heuristic optimization. However, an *"(Almost) No Free Lunch"* ((A)NFL) theorem shows that for each function that can be efficiently optimized by a heuristic search, many other related functions can be constructed where the same heuristic is bad. Consequently, heuristic search methods use some a priori known information, a kind of "ideas," of how to search for good solutions and so they can be successful only for functions that give the appropriate "help."

Theorem 5 *Assume that* **S** *is a randomized search strategy and let* f *be a function,* $f \in \{0, 1\}^n$ *and the output range is* $\{0, 1, 2, \ldots, N - 1\}$. *Then there are at least* $N^{2^{n/3}-1}$ *number of functions, let* g, $g : \{0, 1\} \to \{0, 1, 2, \ldots, N\}$ *in agreement with* f *on all but at most* $2^{n/3}$ *inputs such that* **S** *does find the optimum of* g *within* $2^{n/3}$ *steps with a probability bounded above by* $2^{-n/3}$.

This theorem suggests that heuristic methods cannot succeed in all existing problems. This is because the effectiveness of these techniques is largely based on a good "guess." If this guess is correct, these methods can be very efficient. If not, the search time can reach exponential levels and this constitutes a serious disadvantage of this family of methods.

In [23] Griffiths and Orponen studied optimization strategies for a given finite set of functions. Specifically, they investigated the conditions that need to be satisfied for the functions under consideration in order to have the same performance for a uniform distribution of functions. The result of this research is related to some non-trivial Boolean functions and bounded search algorithms. An important conclusion of this research is that the relationship of NFL theorems and the closed under permutation conditions does not always hold. This happens when we consider functions used to maximize the performance of bounded length searches.

Closing this section, it is worth to mention the contribution of Poli and Graff [44] concerning the NFL theorems and hyper-heuristic techniques. Their conclusions further support the previously referenced works as far as the non-validity of the NFL theorems and the existence of Free Lunches are concerned. The NFL theorems guarantee that this phenomenon happens to hyper-heuristic techniques and high-level hyper-heuristics, if all the problems of interest are closed under permutation. For many real applications the corresponding optimization problems do not satisfy this condition and so, in these cases, there is a Free Lunch for hyper-heuristic techniques. Note that this happens provided that at each level of the search hierarchy the heuristics are evaluated using performance measures that reveal the differences in immediately lower level. The fact that the results of NFL theorems may not hold over heuristic searching techniques does not mean that the existing hyper-heuristic methods are good enough. This may need to be proven and so it requires to be further investigated. Finally, whenever implementation of the NFL theorems is not possible, one should see the opportunity to try finding some new and more powerful, effective, and efficient hyper-heuristic algorithms, including techniques that are based on genetic programming and genetic algorithms.

6 NFL for Supervised Learning

Revisiting his initial work on supervised learning Wolpert in his work entitled: *"The supervised learning no-free-lunch theorems"* [59] analyzes the main issues underlying his theory on NFL. Wolpert criticizes conventional testing methods for supervised learning as they do not account for out-of-sample testing which is more important for the behavior of supervised learning algorithms. Actually, despite any opposite claim it is common practice in established supervised learning approaches to perform testing with test set that overlap training sets. Thus, conventional frameworks are bound with specific application fields of supervised learning and not with the very problems of the domain.

To cope with this inability of conventional frameworks and deal with the off-training-set error he proposes the so-called *Extended Bayesian Framework* (EBF) which besides offering an extension to classical Bayesian analysis it, also, has the major advantage that it encompasses the conventional frameworks. Based on the EBF, Wolpert develops the set of No Free Lunch theorems which *"bound how much one can infer concerning the (off-training-set) generalization error probability distribution without making relatively strong assumptions concerning the real world. They serve as a broad context in which one should view the claims of any supervised learning framework."*

All aspects of supervised learning are modeled by means of probability distributions. Wolpert provides definitions for those points that are ill defined and they are assumed to constitute defaults for conventional approaches which deal with generalization. Hence, according to Wolpert's notation, if

$$d = \{d_X(i), d_Y(i)\}, \quad \forall\, 1 \leqslant i \leqslant m,$$

denotes the training data, "f" is the function giving the probability $P(y \mid x, f) = f_{x,y}$ and "h" is the x-conditioned probability distribution over values y which is produced by the learning algorithm in response to training data d, $P(y|x, h) = h_{x,y}$ then the generalization error function typically used in supervised learning can be expressed by the expectation value $E(C \mid h, f, d)$ for some cost "C" induced by the learning algorithm. So, for the "average misclassification rate error," one may set:

$$E(C \mid h, f, d) = E(C \mid h, f) = \sum_x \pi(x) \left[1 - \delta(f(x), h(x))\right].$$

This is the average number of times across all $x \in X$ that h and f differ relatively to the sampling distribution $\pi(x)$ which produced the training data.

In the sequel, the following two theorems are formulated. These theorems are known as *"No Free Lunch theorems for supervised learning."*

Theorem 6 $E(C \mid d)$ *can be written as a (non-Euclidean) inner product between the distributions* $P(h \mid d)$ *and* $P(f \mid d)$:

$$E(C \mid d) = \sum_{h,f} Er(h, f, d) \, P(h|d) \, P(f|d),$$

where $Er(h, f, d)$ *denotes the error function.*

The following meanings are given by Wolpert to this theorem:

(a) An answer to how well a learning algorithm does on some problem is determined by how "aligned" the algorithm $P(h|d)$ is with the posterior $P(f \mid d)$.
(b) One cannot prove anything regarding how well a particular learning algorithm generalizes as one is, generally, unable to prove that $P(h|d)$ is aligned with $P(f \mid d)$ unless $P(f \mid d)$ has a certain form.

The impossibility to prove that $P(f \mid d)$ has a certain form is formalized by the following theorem.

Theorem 7 *Consider the off-training-set error function. Let "$E_i(\cdot)$" indicate an expectation value evaluated using learning algorithm "i." Then for any two learning algorithms* $P_1(h \mid d)$ *and* $P_2(h \mid d)$, *independent of the sampling distribution*

 (i) *Uniformly averaged over all* f,
    $E_1(C \mid f, m) - E_2(C \mid f, m) = 0$;
 (ii) *Uniformly averaged over all* f, *for any training set* d,
    $E_1(C \mid f, d) - E_2(C \mid f, d) = 0$;
(iii) *Uniformly averaged over all* $P(f)$,
    $E_1(C \mid m) - E_2(C \mid m) = 0$;
(iv) *Uniformly averaged over all* $P(f)$, *for any training set* d,
    $E_1(C \mid d) - E_2(C \mid d) = 0$.

Remark 2 Given that the quantities $E(C \mid d)$, $E(C \mid m)$, $E(C \mid f, d)$, or $E(C \mid f, m)$ denote different measures of risks, the theorem states that for any of these measures any two algorithms on average perform equally well. Actually, Algorithm 1 is superior to Algorithm 2 for as many problems as Algorithm 2 is superior to Algorithm 1.

The examples given by Wolpert are about cross-validation and Bayesian inference. Moreover, some variants of Theorem 7 are presented and the intuitive ideas of Theorem 7 are analyzed. These ideas gave rise to the following important research efforts concerning two critical issues of supervised learning, namely *early stopping* and *cross-validation*.

6.1   No Free Lunch for Early Stopping

Iterative methods, such as gradient descent, train a learner by updating its free parameters in order to make it better fit the training data and improve the performance of the learner on data outside the training set. Up to some point this is a successful task but beyond that point further training leads to over-fitting the training data while failing to deal with out-of-sample data, thus, increasing the generalization error of the learner. Regularization techniques including early stopping are used to avoid over-fitting.

The *"early stopping"* provides rules on how to conduct training and when to stop iterations in order to avoid over-fitting. In machine learning the early stopping has been used in many contexts and has been supported with various mathematical tools. A well-known and widely used technique is to guide validation of a training procedure with early stopping by monitoring the increase of the generalization error on validation data.

In [7] Cataltepe et al. aim at bringing the idea of NFL into the framework of early stopping. The method of choosing a model using the early stopping approach relies on a uniform selection of the model among the models giving the same training error. This approach is claimed to be similar to the *"Gibbs algorithm."* The uniform probability of selection around the training error minimum is equivalent to the isotropic distributions of Amari et al. [3], while it differs from this work as it does not assume a very large number of training examples. In addition to general linear models in [7] it is presumed that the probability of selection of models is symmetric only around the training error minimum.

This symmetry hypothesis is a weaker requirement than uniformity. The authors analyze early stopping for some training error minimum. If the training set constitutes all the information that one has about the target, then one should minimize the training error as much as possible to achieve lower generalization error. Moreover, the authors demonstrate that when additional information is available, early stopping can help.

6.2   No Free Lunch for Cross-Validation

In machine learning and, generally, in statistical learning theory, *"cross-validation"* is a model evaluation method used when a predictive modeling procedure or any learner is asked to make new predictions for data it has not already seen. This data constitutes the model validation set. Therefore, instead of using mathematical analysis cross-validation is a generally applicable method used to assess the performance of a model. Specific methods of cross-validation can be either of *"exhaustive"* (such as leave p-out, leave-one-out) or *"non-exhaustive"* type (such as k-fold, hold out, repeated random sub-sampling) and they are able to give meaningful results provided that the training set and the validation set are drawn from the same population, i.e., the same distribution.

Cross-validation is a statistical technique which constitutes an objective approach to compare different learning procedures as it does not rely on in-sample error rates. Thus, it was long widely believed that it can be successful, despite of the prior knowledge available on the problem at hand. Zhu and Rohwer [68] provide a numerical counter-example which, despite the fact that it is an artificial one, constitutes a minimal proof that cross-validation is not a *"universally beneficial method."* The problem consists in selecting the unbiased estimator of the expectation of a Gaussian distribution between two estimators, namely: (a) an unbiased and (b) a highly biased one. The authors apply the leave-one-out scheme and make an attempt to show that this method is inefficient even in small problems. Hence, cross-validation cannot defy the theoretical result of the NFL theorem, that is, *"no algorithm can be good for any arbitrary prior."*

Moreover, the authors carry out further experiments and give a detailed analysis with the aim to provide answers to any criticism against the main issue tackled by the paper which is *"as with any other algorithm, cross-validation and, in this sense, a number of other approaches such as bootstrap cannot solve equally good any kind of problem."* Hence, if some prior knowledge is used for an algorithm, then this should be communicated to any interested user so that he can decide whether to use it or not.

Goutte published a more elaborated approach on this matter in his work [22] entitled *"Note on free lunches and cross-validation."* In this paper the aforementioned approach of Zhu and Rohwer on cross-validation and NFL theorem is revisited by further elaborating on the numerical example. The author, also, applies the leave-one-out and the m-fold cross-validation schemes on the numerical result used by Zhu and Rohwer and performs a more detailed mathematical description. Analysis of the results obtained supports the argument that there is *"No Free Lunch for cross-validation"* and though the method is not the best approach for evaluating performance of learners, however, it is capable to give very good results in a number of practical situations.

Further to the above research, Rivals and Personnaz [46] took over the work of Goutte and by using probabilistic analysis they applied leave-one-out cross-validation on measures of model quality. The leave-one-out scores obtained show that the conclusions of Goutte are optimistic as they deal with a trivial problem for which any reasonable method is not prone to make a wrong choice. In addition, a comparison between leave-one-out cross-validation and statistical tests for the selection of linear models is performed. The numerical results obtained by a specific illustrative example show that for linear estimators with large number of samples, leave-one-out cross-validation does not perform well as compared to statistical tests. This leads to the conclusion that it is unlikely that this method is able to perform well in the case of nonlinear estimators such as neural networks. Hence, an important result is stressed, that is, statistical tests should be preferred to leave-one-out cross-validation *"provided that the (linear or nonlinear) model has the properties required for the statistical tests to be valid."*

6.3 Real-World Machine Learning Classification and No Free Lunch Theorems: An Experimental Approach

The majority of the research concerning NFL and supervised learning seems to be more or less theoretical. Unlike the previously reported work Gómez and Rojas published in [21] the results of a number of machine learning experiments with the aim to help understanding the impact of NFL on real-world problems. At the same time the authors attempted to provide sufficient experimental evidence on the validity of NFL.

The set of machine learning algorithms used in these experiments comprise:

(a) Naive Bayes classifiers,
(b) C4.5 decision trees,
(c) Neural networks,
(d) k-nearest neighbors classifiers,
(e) Random C4.5 forest,
(f) AdaBoost.M1,
(g) Stacking.

The performance of these approaches was examined in terms of average accuracy over six data sets taken from the UCI machine learning repository. These data sets are: "Audiology," "Column," "Breast cancer," "Multiple features (Fourier)," "German credit," and "Nursery." To a great extent, the results obtained are consistent with previous research. On the other hand, according to the NFL theorem the tested algorithms should expose the same degree of accuracy. However, this is valid when a sufficiently large number of data sets are available. The authors underline that some common assumptions pertain the data sets. These common assumptions concern the Occam's razor and the independent identical distribution of the samples as well as, mainly, the data-dependent structural properties found in the data sets, that is, determinism and the Pareto principle. Based on these last properties they explain the peculiarities of the data sets and the results concerning the accuracy. Then, it is clear that not all the algorithms perform equally well on all problems.

In addition to the above, the authors perform a number of experiments using kernel machines and especially support vector machines (SVM) as well as deep learning networks. The results obtained show that SVM outperform the other learning algorithms while the performance of deep learning on these small and relatively simple problems is disappointing. In fact while these architectures are designed to handle complex data sets which have inherent abstraction layers they seem to be incapable to cope with simpler data sets with possibly lower data abstraction. This shows that NFL applies even in the case of deep learning which is also subject to limitations as other machine learning algorithms.

In terms of conclusion the authors state that: *"While evaluating the average accuracy ranking for the six data sets, they noticed the effect of the NFL theorem and how assumptions are key to performance."* Comparing with similar research work they conclude that: *"the data and its pre-processing are as important as, if not more so than, the algorithm itself in determining the quality of the model. Data visualization or statistical techniques such as feature selection can be crucial to provide a better fit and obtain simpler and better models."*

7 Synopsis and Concluding Remarks

In this paper we surveyed some of the most sound research works concerning No Free Lunch (NFL) theorems and their results in search, optimization, and supervised learning. Starting from the earlier work of David H. Wolpert, where the essential concepts underpinning NFL theorems were defined, we went through the research efforts that contributed to the formulation of the most relevant frameworks for applying NFL theorems. Moreover, we presented those research works which show when NFL theorems do not hold and so there are Free Lunches, i.e., algorithms that significantly outperform other algorithms on specific classes of problems. One of the objectives set for this survey was to make clear which are the hypotheses and the restrictions for applying NFL results, or on the contrary, to pinpoint the conditions under which there are Free Lunches, as defined by researchers in their respective papers.

One of the most relevant conclusion is that important research needs to be carried out in order to delineate those classes of problems for which NFL theorems apply and those for which they don't. NFL theorems do not put any obstacle on continuing research for developing more efficient algorithms which apply to even larger classes of problems. They just seem to make clear that there are limits for these algorithms. Finally, this does not mean that for some specific problems one is not able to design an algorithm performing better than its competitors.

Closing this review we hope that this work will assist all those who are interested in NFL theorems.

Acknowledgements S.-A. N. Alexandropoulos is supported by Greece and the European Union (European Social Fund-ESF) through the Operational Programme "Human Resources Development, Education and Lifelong Learning" in the context of the project "Strengthening Human Resources Research Potential via Doctorate Research" (MIS-5000432), implemented by the State Scholarships Foundation (IKY). P. M. Pardalos is supported by the Paul and Heidi Brown Preeminent Professorship at ISE (University of Florida, USA), and a Humboldt Research Award (Germany).

References

1. Al-Rifaie, M.M., Bishop, J.M.: Swarmic paintings and colour attention. In: International Conference on Evolutionary and Biologically Inspired Music and Art, pp. 97–108. Springer, Berlin (2013)
2. Al-Rifaie, M.M., Bishop, J.M., Caines, S.: Creativity and autonomy in swarm intelligence systems. Cogn. Comput. **4**(3), 320–331 (2012)

3. Amari, S., Murata, N., Muller, K.R., Finke, M., Yang, H.H.: Asymptotic statistical theory of overtraining and cross-validation. IEEE Trans. Neural Netw. **8**(5), 985–996 (1997)
4. Auger, A., Teytaud, O.: Continuous lunches are free plus the design of optimal optimization algorithms. Algorithmica **57**(1), 121–146 (2010)
5. Auger, A., Schoenauer, M., Teytaud, O.: Local and global order 3/2 convergence of a surrogate evolutionary algorithm. In: Proceedings of the 7th Annual Conference on Genetic and Evolutionary Computation, pp. 857–864. ACM, New York (2005)
6. Bellman, R.: Dynamic Programming. Princeton University Press, Princeton (1957)
7. Cataltepe, Z., Abu-Mostafa, Y.S., Magdon-Ismail, M.: No free lunch for early stopping. Neural Comput. **11**(4), 995–1009 (1999)
8. Ciucu, F., Schmitt, J.: Perspectives on network calculus: no free lunch, but still good value. ACM SIGCOMM Comput. Commun. Rev. **42**(4), 311–322 (2012)
9. Corne, D., Knowles, J.: Some multiobjective optimizers are better than others. In: IEEE Congress on Evolutionary Computation (CEC 2003), vol. 4, pp. 2506–2512. IEEE, Piscataway (2003)
10. Dembski, W.A.: No Free Lunch: Why Specified Complexity Cannot be Purchased Without Intelligence. Rowman & Littlefield, Langham (2006).
11. Dorigo, M., Birattari, M.: Ant colony optimization. In: Encyclopedia of Machine Learning, pp. 36–39. Springer, Boston (2011)
12. Drettakis, G., Roussou, M., Reche, A., Tsingos, N.: Design and evaluation of a real-world virtual environment for architecture and urban planning. Presence Teleop. Virt. **16**(3), 318–332 (2007)
13. Droste, S., Jansen, T., Wegener, I.: Optimization with randomized search heuristics – the (A)NFL theorem, realistic scenarios, and difficult functions. Theor. Comput. Sci. **287**(1), 131–144 (2002)
14. Eberhart, R., Kennedy, J.: A new optimizer using particle swarm theory. In: Proceedings of the IEEE Sixth International Symposium on Micro Machine and Human Science, 1995, MHS'95, pp. 39–43. IEEE, Piscataway (1995)
15. Epitropakis, M.G., Plagianakos, V.P., Vrahatis, M.N.: Evolutionary adaptation of the differential evolution control parameters. In: Proceedings of the IEEE Congress on Evolutionary Computation, 2009, CEC'09, pp. 1359–1366. IEEE, Piscataway (2009)
16. Epitropakis, M.G., Tasoulis, D.K., Pavlidis, N.G., Plagianakos, V.P., Vrahatis, M.N.: Enhancing differential evolution utilizing proximity-based mutation operators. IEEE Trans. Evol. Comput. **15**(1), 99–119 (2011)
17. Epitropakis, M.G., Plagianakos, V.P., Vrahatis, M.N.: Evolving cognitive and social experience in particle swarm optimization through differential evolution: a hybrid approach. Inf. Sci. **216**, 50–92 (2012)
18. Ficici, S.G.: Solution Concepts in Coevolutionary Algorithms. PhD thesis, Brandeis University Waltham, Waltham (2004)
19. Floudas, C.A., Pardalos, P.M.: Encyclopedia of Optimization. Springer Science & Business Media B.V., Dordrecht (2008)
20. Georgiou, V.L., Malefaki, S., Parsopoulos, K.E., Alevizos, Ph.D., Vrahatis, M.N.: Expeditive extensions of evolutionary Bayesian probabilistic neural networks. In: Third International Conference on Learning and Intelligent Optimization (LION3 2009). Lecture Notes in Computer Science, vol. 5851, pp. 30–44. Springer, Berlin (2009)
21. Gómez, D., Rojas, A.: An empirical overview of the no free lunch theorem and its effect on real-world machine learning classification. Neural Comput. **28**(1), 216–228 (2015)
22. Goutte, C.: Note on free lunches and cross-validation. Neural Comput. **9**(6), 1245–1249 (1997)
23. Griffiths, E.G., Orponen, P.: Optimization, block designs and no free lunch theorems. Inf. Process. Lett. **94**(2), 55–61 (2005)
24. Ho, Y.C.: The no free lunch theorem and the human-machine interface. IEEE Control. Syst. **19**(3), 8–10 (1999)

25. Hopkins, D.A., Thomas, M.: Neural network and regression methods demonstrated in the design optimization of a subsonic aircraft. Structural Mechanics and Dynamics Branch 2002 Annual Report, p. 25 (2003)
26. Hume, D. (Introduction by Mossner, E.C.): A Treatise of Human Nature. Classics Series. Penguin Books Limited, London (1986)
27. Hume, D.: A Treatise of Human Nature. The Floating Press Ltd., Auckland (2009). First published in 1740
28. Karaboga, D., Basturk, B.: A powerful and efficient algorithm for numerical function optimization: artificial bee colony (ABC) algorithm. J. Glob. Optim. **39**(3), 459–471 (2007)
29. Kimbrough, S.O., Koehler, G.J., Lu, M., Wood, D.H.: On a feasible–infeasible two-population (FI-2Pop) genetic algorithm for constrained optimization: distance tracing and no free lunch. Eur. J. Oper. Res. **190**(2), 310–327 (2008)
30. Kleijnen, J.P.C.: Sensitivity analysis of simulation experiments: regression analysis and statistical design. Math. Comput. Simul. **34**(3–4), 297–315 (1992)
31. Kocsis, L., Szepesvari, C.: Bandit-based Monte-Carlo planning. In: European Conference on Machine Learning (ECML 2006). Lecture Notes in Computer Science, vol. 4212, pp. 282–293. Springer, Berlin (2006)
32. Köppen M.: Some technical remarks on the proof of the no free lunch theorem. In: Proceedings of the Fifth Joint Conference on Information Sciences (JCIS), vol. 1, pp. 1020–1024. Atlantic City (2000)
33. Köppen, M., Wolpert, D.H., Macready, W.G.: Remarks on a recent paper on the "No Free Lunch" theorems. IEEE Trans. Evol. Comput. **5**(3), 295–296 (2001)
34. Laskari, E.C., Parsopoulos, K.E., Vrahatis, M.N.: Utilizing evolutionary operators in global optimization with dynamic search trajectories. Numer. Algorithms **34**(2–4), 393–403 (2003)
35. Macready, W.G., Wolpert, D.H.: What makes an optimization problem hard? Complexity **1**(5), 40–46 (1996)
36. Marshall, J.A.R., Hinton, T.G.: Beyond no free lunch: Realistic algorithms for arbitrary problem classes. In: IEEE Congress on Evolutionary Computation, pp. 1–6. IEEE, Piscataway (2010)
37. Orr, H.A.: Review of no free lunch by William A Dembski. Boston Review. Available on-line at http://bostonreview.net/BR27, 3 (2002)
38. Parsopoulos, K.E., Vrahatis, M.N.: Recent approaches to global optimization problems through particle swarm optimization. Nat. Comput. **1**(2–3), 235–306 (2002)
39. Parsopoulos, K.E., Vrahatis, M.N.: On the computation of all global minimizers through particle swarm optimization. IEEE Trans. Evol. Comput. **8**(3), 211–224 (2004)
40. Parsopoulos, K.E., Vrahatis, M.N.: Parameter selection and adaptation in unified particle swarm optimization. Math. Comput. Model. **46**(1–2), 198–213 (2007)
41. Parsopoulos, K.E., Vrahatis, M.N.: Particle Swarm Optimization and Intelligence: Advances and Applications. Information Science Publishing (IGI Global), Hershey (2010)
42. Perakh, M.: The No Free Lunch Theorems and Their Application to Evolutionary Algorithms (2003)
43. Petalas, Y.G., Parsopoulos, K.E., Vrahatis, M.N.: Memetic particle swarm optimization. Ann. Oper. Res. **156**(1), 99–127 (2007)
44. Poli, R., Graff, M.: There is a free lunch for hyper-heuristics, genetic programming and computer scientists. In: Proceedings of the 12th European Conference on Genetic Programming, EuroGP '09, pp. 195–207. Springer, Berlin (2009)
45. Poli, R., Graff, M., McPhee, N.F.: Free lunches for function and program induction. In: Proceedings of the Tenth ACM SIGEVO Workshop on Foundations of Genetic Algorithms, FOGA '09, pp. 183–194. ACM, New York (2009)
46. Rivals, I., Personnaz, L.: On cross validation for model selection. Neural Comput. **11**(4), 863–870 (1999)
47. Rosenberg, L.B.: Human swarms, a real-time paradigm for collective intelligence. Collective Intelligence (2015)

48. Schumacher, C., Vose, M.D., Whitley, L.D.: The no free lunch and problem description length. In: Proceedings of the 3rd Annual Conference on Genetic and Evolutionary Computation, pp. 565–570. Morgan Kaufmann Publishers Inc., Burlington (2001)
49. Service, T.C., Tauritz, D.R.: A no-free-lunch framework for coevolution. In: Proceedings of the 10th Annual Conference on Genetic and Evolutionary Computation, pp. 371–378. ACM, Piscataway (2008)
50. Sotiropoulos, D.G., Stavropoulos, E.C., Vrahatis, M.N.: A new hybrid genetic algorithm for global optimization. Nonlinear Anal. Theory Methods Appl. **30**(7), 4529–4538 (1997)
51. Teytaud, O., Flory, S.: Upper confidence trees with short term partial information. In: European Conference on the Applications of Evolutionary Computation, pp. 153–162. Springer, Berlin (2011)
52. Thalmann, D.: Crowd Simulation. Wiley Online Library (2007)
53. Van Grieken, M.: Optimisation pour l'apprentissage et apprentissage pour l'optimisation. PhD thesis, Université Paul Sabatier-Toulouse III, Toulouse (2004)
54. Vanaret, C., Gallard, F., Martins, J.: On the consequences of the "No Free Lunch" theorem for optimization on the choice of an appropriate MDO architecture. In: 18th AIAA/ISSMO Multidisciplinary Analysis and Optimization Conference, pp. 3148 (2017)
55. Wolpert, D.H.: On the connection between in-sample testing and generalization error. Complex Syst. **6**(1), 47–94 (1992)
56. Wolpert, D.H.: The lack of a priori distinctions between learning algorithms. Neural Comput. **8**(7), 1341–1390 (1996)
57. Wolpert, D.H.: The existence of a priori distinctions between learning algorithms. Neural Comput. **8**(7), 1391–1420 (1996)
58. Wolpert, D.H.: The supervised learning no-free-lunch theorems. In: Soft Computing and Industry, pp. 25–42. Springer, London (2002)
59. Wolpert, D.H.: The Supervised Learning No-Free-Lunch Theorems, pp. 25–42. Springer, London (2002)
60. Wolpert, D.H.: What the no free lunch theorems really mean; how to improve search algorithms. SFI working paper: 2012–10-017. Santa Fe Institute, Santa Fe (2012)
61. Wolpert, D.H., Macready, W.G.: No Free Lunch Theorems for Search. Tech. Rep. SFI-TR-95-02-010. Santa Fe Institute, Santa Fe (1995)
62. Wolpert, D.H., Macready, W.G.: No free lunch theorems for optimization. IEEE Trans. Evol. Comput. **1**(1), 67–82 (1997)
63. Wolpert, D. H., Macready, W.G.: Coevolutionary free lunches. IEEE Trans. Evol. Comput. **9**(6), 721–735 (2005)
64. Yang, X.S.: Firefly algorithm, stochastic test functions and design optimization. Int. J. Bio-Inspired Comput. **2**(2), 78–84 (2010)
65. Yang, X.S.: A new metaheuristic bat-inspired algorithm. In: Nature Inspired Cooperative Strategies for Optimization (NICSO 2010), pp. 65–74. Springer, Berlin (2010)
66. Yang, X.S.: Swarm-based metaheuristic algorithms and no-free-lunch theorems. In: Theory and New Applications of Swarm Intelligence. InTech, London (2012)
67. Yang, X.S., Deb, S.: Cuckoo search via Lévy flights. In: Proceedings of the World Congress on Nature & Biologically Inspired Computing, 2009, NaBIC 2009. pp. 210–214. IEEE, Piscataway (2009)
68. Zhu, H., Rohwer, R.: No free lunch for cross-validation. Neural Comput. **8**(7), 1421–1426 (1996)

Piecewise Convex–Concave Approximation in the Minimax Norm

Michael P. Cullinan

Abstract Suppose that $\mathbf{f} \in \mathbb{R}^n$ is a vector of n error-contaminated measurements of n smooth function values measured at distinct, strictly ascending abscissæ. The following projective technique is proposed for obtaining a vector of smooth approximations to these values. Find $\mathbf{y}$ minimizing $\|\mathbf{y} - \mathbf{f}\|_\infty$ subject to the constraints that the consecutive second-order divided differences of the components of $\mathbf{y}$ change sign at most q times. This optimization problem (which is also of general geometrical interest) does not suffer from the disadvantage of the existence of purely local minima and allows a solution to be constructed in only $O(nq \log n)$ operations. A new algorithm for doing this is developed and its effectiveness is proved. Some results of applying it to undulating and peaky data are presented, showing that it is fast and can give very good results, particularly for large densely packed data, even when the errors are quite large.

1 Introduction

This paper proposes a new fast one-dimensional data-smoothing algorithm which gives particularly good results for large densely packed data with small errors, for which there are no very satisfactory methods currently available, and which is also of geometrical interest.

Suppose a set of one-dimensional observations are known to be measurements of smooth quantities contaminated by errors. A method is then needed to get a smooth set of points while respecting the observations as much as possible. One method is to make the least change to the observations, measured by a suitable norm, in order to achieve a prescribed definition of smoothness. The data smoothing method of Cullinan and Powell [3] proposes defining smoothness as the consecutive divided differences of the points of a prescribed order r having at most a prescribed number

M. P. Cullinan (✉)
Maryvale Institute, Birmingham, UK

© Springer Nature Switzerland AG 2019
I. C. Demetriou, P. M. Pardalos (eds.), *Approximation and Optimization*,
Springer Optimization and Its Applications 145,
https://doi.org/10.1007/978-3-030-12767-1_6

q of sign changes. This is an economical and sensitive test for smoothness because normally data values of a smooth function will have very few sign changes, whereas if even one error is introduced, it will typically cause k sign changes in the kth order divided differences of the contaminated data (see, for example, Hildebrand [7]).

If the observations f_j, at strictly ascending abscissæ x_j, for $1 \leq j \leq n$, are regarded as the components of a vector $\mathbf{f} \in \mathbb{R}^n$ and the function $F :$ $\mathbb{R}^n \to \mathbb{R}$ is defined through the chosen norm by $F(\mathbf{v}) = \|\mathbf{v} - \mathbf{f}\|$, then the data smoothing problem becomes the constrained minimization of F. This approach has several advantages. There is no need to choose (more or less arbitrarily) a set of approximating functions, indeed the data are treated simply as the set of finite points which they are rather than as coming from any underlying function. The method is projective or invariant in the sense that it leaves smoothed points unaltered. It depends on two integer parameters which will usually take only a small range of possible values, rather than requiring too arbitrary a choice of parameters. It may be possible to choose likely values of q and r by inspection of the data. The choice of norm can sometimes be suggested by the kind of errors expected, if this is known. For example, the ℓ_1 norm is a good choice if a few very large errors are expected, whereas the ℓ_∞ norm might be expected to deal well with a large number of small errors. There is also the possibility that the algorithms to implement the method may be very fast.

The main difficulty in implementing this method is that when $q \geq 1$, the possible existence of purely local minima of F makes the construction of an efficient algorithm very difficult. This has been done for the ℓ_2 norm for $r \leq 2$—see, for example, Demetriou and Powell [5] and Demetriou [4]. The author has dealt with the case $q = 0$ and arbitrary r for the ℓ_2 norm (Cullinan [2]).

It was claimed in Cullinan and Powell [3] that when the ℓ_∞ norm is chosen and $r = 2$, all the local minima of F are global and a best approximation can be constructed in $O(nq)$ operations—rather than n^q operations as might be expected. An algorithm for doing this was outlined. These claims were proved by Cullinan [1] which also considered the case $r = 1$. It was shown that in these cases the minimum value of F is determined by $q + r + 1$ of the data, and a modified algorithm for the case $r = 2$ was developed which is believed to be better than that outlined in Cullinan and Powell [3].

A refined version of this new algorithm will now be presented in Section 2 and its effectiveness will be proved. Section 3 will then describe the results of some tests of this method which show that it is a very cheap and efficient way of filtering noise but can be prone to end errors.

2   The Algorithm

This section will construct a best ℓ_∞ approximation to a vector $\mathbf{f} \in \mathbb{R}^n$ with not more than q sign changes in its second divided differences. More precisely let $\mathbf{v} \in \mathbb{R}^n$ with $x_1 < x_2 < \ldots < x_n$ and

$$F(\mathbf{v}) = \|\mathbf{v} - \mathbf{f}\|_\infty \tag{1}$$

$$c_{ijk}(\mathbf{v}) = \frac{1}{x_k - x_i} \left(\frac{v_k - v_j}{x_k - x_j} - \frac{v_j - v_i}{x_j - x_i} \right), \tag{2}$$

$$c_i(\mathbf{v}) = c_{i,i+1,i+2}(\mathbf{v}). \tag{3}$$

The set of feasible points $Y_q \subset \mathbb{R}^n$ is defined as the set of all vectors $\mathbf{v} \in \mathbb{R}^n$ for which the signs of the successive elements of the sequence $1, c_1(\mathbf{v}), \ldots, c_{n-2}(\mathbf{v})$ change at most q times, and the problem is then to develop an algorithm to minimize F over Y_q.

Define h_q to be the value of the best approximation to F over Y_q. The solution depends on the fact that h_q is determined by $q + 3$ of the data. Since a best ℓ_∞ approximation is not unique, there is some choice of which one to construct. The one chosen, $\mathbf{y}$, has the following property: $y_1 = f_1 + h_q$, $y_n = f_n + (-1)^q h_q$, and for any j with $2 \leq j \leq n - 1$,

$$\text{if } \pm c_{j-1}(\mathbf{y}) > 0 \text{ then } y_j = f_j \pm h_q.$$

The vector $\mathbf{y}$ is then determined from h_q, from the set of indices i where $c_{i-1}(\mathbf{y}) \neq 0$, and from the ranges where the divided differences do not change sign.

The method by which the best approximation is constructed and the proof of the effectiveness of the algorithm that constructs it are best understood by first considering the cases $q = 0, 1$, and 2 in detail and giving algorithms for the construction of a best approximation in each case. Once this has been done it is much easier to understand the somewhat complicated bookkeeping required for the general algorithm.

When $q = 0$, this best approximation is formed from the ordinates of the points on the lower part of the boundary of the convex hull of the points (x_j, f_j), for $1 \leq j \leq n$, (the graph of the data in the plane) by increasing these ordinates by an amount h.

When $q = 1$, there exist integers s and t such that $1 \leq s \leq t \leq n$, and $y_1, \ldots, y_s$ are the ordinates on the lower part of the boundary of the convex hull of the data $f_1, \ldots, f_s$ increased an amount h; $y_t, \ldots, y_n$ are the ordinates on the upper part of the boundary of the concave hull of the data $f_t, \ldots, f_n$ decreased an amount h; and if $s < j < t$, y_j lies on the straight line joining (x_s, y_s) to (x_t, y_t). The best approximation therefore consists of a convex piece and a concave piece joined where necessary by a straight line.

The best approximation over Y_q consists of $q + 1$ alternately raised pieces of lower boundaries of convex hulls and lowered pieces of upper boundaries of concave hulls joined where necessary by straight lines. These pieces are built up recursively from those of the best approximation over Y_{q-2}.

The points on the upper or lower part of the boundary of the convex hull of the graph of a range of the data each lie on a convex polygon and are determined from its vertices. The algorithms to be described construct sets of the indices of these vertices and the value of a best approximation. The best approximation vector is then constructed by linear interpolation.

Before considering the cases $q = 0$, 1, and 2, an important preliminary result will be established. It is a tool that helps to show that the vectors constructed by the algorithms are optimal. The value h_q of the best approximation over Y_q will be found by the algorithm, together with a vector $\mathbf{y} \in Y_q$ such that $F(\mathbf{y}) = h_q$. To show that $\mathbf{y}$ is optimal, a set K of $q + 3$ indices will be constructed such that if $\mathbf{v}$ is any vector in $\mathbb{R}^n$ such that $F(\mathbf{v}) < h_q$, then the consecutive second divided differences of the components v_k, for $k \in K$, change sign q times starting with a negative one. It will then be inferred that $\mathbf{v} \notin Y_q$. In order to make this inference it must be shown that the consecutive divided differences of *all* the components of $\mathbf{v}$ have at least as many sign changes as those of the components with indices in K. This result will now be proved.

Theorem 1 *Let $K \subseteq \{1, \ldots, n\}$ and let $\mathbf{v} \in \mathbb{R}^n$ be any vector such that the second divided differences of the v_k, for $k \in K$, change sign q times. Then the divided differences of all the components of $\mathbf{v}$ change sign at least q times.*

Proof Firstly, suppose that K is formed by deleting one element j from $\{1, \ldots, n\}$ and that $3 \le j \le n - 2$. Let c_{j-2}, c_{j-1}, c_j be defined from $\mathbf{v}$ by (2) and (3), and let $c'_{j-2} = c_{j-2,j-1,j+1}(\mathbf{v})$ and $c'_{j-1} = c_{j-1,j+1,j+2}(\mathbf{v})$ denote the new divided differences that result from deleting j. Manipulation of (2) yields the equations

$$c'_{j-2} = \frac{x_j - x_{j-2}}{x_{j+1} - x_{j-2}} c_{j-2} + \frac{x_{j+1} - x_j}{x_{j+1} - x_{j-2}} c_{j-1}$$

and

$$c'_{j-1} = \frac{x_{j+2} - x_j}{x_{j+2} - x_{j-1}} c_j + \frac{x_j - x_{j-1}}{x_{j+2} - x_{j-1}} c_{j-1},$$

so that c'_{j-2} lies between c_{j-2} and c_{j-1} and c'_{j-1} lies between c_{j-1} and c_j. It follows that the number of sign changes in the sequence

$$\ldots, c_{j-3}, c_{j-2}, c'_{j-2}, c_{j-1}, c'_{j-1}, c_j, c_{j+1}, \ldots, \tag{4}$$

is the same as that in the sequence

$$\ldots, c_{j-3}, c_{j-2}, c_{j-1}, c_j, c_{j+1} \ldots,$$

and hence that deleting c_{j-2}, c_{j-1}, c_j from (4) cannot increase the number of sign changes. The same argument covers the cases $j = 2$ and $j = n-1$, and when $j = 1$ or $j = n$ this result is immediate.

Repeated application of this result as elements j of the set $\{1, \ldots, n\}\backslash K$ are successively deleted from $\{1, \ldots, n\}$ then proves the theorem.   □

This theorem has a corollary that is important in showing that the approximations to be produced are optimal.

Corollary 1 *If $i \leq k - 2$ and all the divided differences $c_j(\mathbf{v}), i < j < k$, are non-negative (or non-positive), and if $i \leq r < s < t \leq k$, then $c_{rst}(\mathbf{v})$ is also non-negative (or non-positive).*

Theorem 1 is crucial to the effectiveness of the algorithm because it allows the explicit construction of global minima determined by $q + 3$ of the data. One reason that this is possible is that, as mentioned above, the set of best approximations is connected. This was proved in Cullinan [1], where it was also shown that there is no analogous result for higher order divided differences, and so no immediate generalization of the methods of this paper to such cases.

2.1 The Case $q = 0$

When $q = 0$, the required solution minimizes (2) over Y_0 and is called a best convex approximation to $\mathbf{f}$. The particular one, $\mathbf{y}^0$, that will be constructed here was first produced by Ubhaya [11]. It will also be convenient to construct a best *concave* approximation to data.

The best convex approximation is determined from the vertices of the lower part of the boundary of the convex hull of the graph of the points in the plane, and the best concave approximation from the vertices of the upper part. These vertices are each specified by sets of indices that can be constructed in $O(n)$ operations by Algorithm 1 below. Much use will be made of this algorithm in the cases where $q \geq 1$ and so it is convenient to apply the construction to a general range of the data and to describe it in terms of sets of indices. Accordingly, define a *range* $[r, s] = \{j : r \leq j \leq s\}$, and a *vertex set* of $[r, s]$ to be any set I such that $\{r, s\} \subseteq I \subseteq [r, s]$. Given a vertex set I of a range $[r, s]$ and also quantities $v_i, i \in I$, define the *gradients*

$$g_{ik}(\mathbf{v}) = \frac{v_k - v_i}{x_k - x_i} \qquad \text{for } i \neq k.$$

A vertex set naturally generates a complete interpolating vector by 'joining the dots' according to the following procedure. Given any integer $j : 1 \leq j \leq n$, define the *neighbours* of j in I by

$$j^+(I) = \min_{i \in I}\{i > j\} \text{ when } j < s$$

$$j^-(I) = \max_{i \in I}\{i < j\} \text{ when } j > r,$$

and, for extrapolation,

$$j^+(I) = s^-(I) \text{ when } j \geq s$$

$$j^-(I) = r^+(I) \text{ when } j \leq r.$$

The *interpolant* $\mathbf{v}(I)$ can now be defined by

$$v_j(I) = v_j \qquad \text{for } j \in I \tag{5}$$

$$v_j(I) = v_{j^-(I)} + g_{j^-(I)\,j^+(I)}(\mathbf{f})(x_j - x_{j^-(I)}) \qquad \text{for all } j \notin I. \tag{6}$$

When $I = \{p, q\}$ it is convenient to write $v_j(I)$ as $v_j(p, q)$, etc.

The two cases of convex and concave approximations are handled using the sign variable σ, where $\sigma = +$ for the convex case and $\sigma = -$ for the concave case. The *convex* and *concave optimal vertex sets* $I^+(r, s)$ and $I^-(r, s)$ are then constructed by systematic deletion as follows.

Algorithm 1 *To find $I^\sigma(r, s)$ when $r \leq s - 2$.*

Step 1.   Set $I := [r, s]$ and $i := r$, $j := r + 1$, $k := r + 2$.
Step 2.   Evaluate $c := c_{ijk}(\mathbf{f})$. If $\sigma c > 0$: go to Step 5.
Step 3.   Delete j from I. If $i = r$: go to Step 5.
Step 4.   Set $j := i$, and $i := i^-(I)$. Go to Step 2.
Step 5.   If $k = s$: set $I^\sigma(r, s) = I$ and **stop.**
          Otherwise: set $i := j$, $j := k$, and $k := k^+(I)$. Go to Step 2.

The *price* of making a convex/concave approximation in the range $[r, s]$ is given from the optimal vertex sets by

$$h^\sigma(r, s) = \tfrac{1}{2} \max_{j \in [r,s]} \sigma(f_j - f_j(I^\sigma(r, s))). \tag{7}$$

The required best approximation $\mathbf{y}^0$ to all the data is then given by

$$y_j^0 = f_j(I) + h, \tag{8}$$

where $I = I^+(1, n)$ and $h = h^+(1, n)$. Some elements of the construction of $\mathbf{y}^0$ are illustrated in Figure 1. Note that the deletion of indices is such that if k is an intermediate element of $I^\sigma(r, s)$ then $c_{k^-,k,k^+}(\mathbf{f})$ is strictly non-zero. For example, if the data are collinear, then $I^\sigma(r, s) = \{r, s\}$.

It is also worth noting here that the prices and constraints are linearly related. Given indices i, j, k and quantities v_i, v_j, v_k then a simple calculation shows that

$$v_j - v_j(i, k) = \frac{-1}{(x_j - x_i)(x_k - x_j)} c_{ijk}(\mathbf{v}). \tag{9}$$

Theorem 2 *Let $n \geq 3$, and let $h_0 = h^+(1, n)$ and $\mathbf{y}^0$ be defined from (7) and (8) using Algorithm 1, (5) and (6). Then $\mathbf{y}^0 \in Y_0$ and $F(\mathbf{y}^0) = h_0 = \inf F(Y_0)$.*

Proof All the proofs of effectiveness of the Algorithms in this paper will follow the same lines. First it will be shown that a well-defined solution vector is produced,

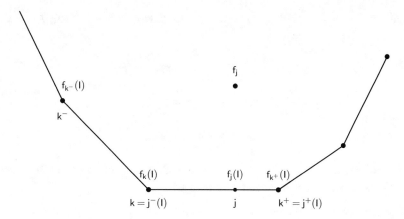

Fig. 1 Consecutive elements k^-, k, k^+ of $I = I^+(1, n)$ and the construction of $f_j(I)$

then that the solution vector is feasible, and then that it is optimal because no vector with a lower value of F can be feasible.

Thus the first remark is that Algorithm 1 does produce a well-defined vertex set I of $[r, s]$ from which the quantities $f_j(I)$ are also well defined for all j, so that h_0 and $\mathbf{y}^0$ are also well defined.

The proof that the points $(x_j, f_j(I))$, $1 \le j \le n$, lie on the lower part of the convex hull of the data is in Ubhaya [11]. The components $f_j(I)$ can be defined as those that are maximal subject to the inequalities

$$f_j(I) \le f_j, \qquad \text{for } 1 \le j \le n, \tag{10}$$

and

$$c_i(\mathbf{f}(I)) \ge 0, \qquad \text{for } 1 \le i \le n - 2, \tag{11}$$

i.e., $\mathbf{f}(I) \in Y_0$. It follows that $\mathbf{y}^0 \in Y_0$ and from (7) and (10) that $F(\mathbf{y}^0) = h$.

It remains to prove that $\mathbf{y}^0$ is optimal. If $h = 0$, $\mathbf{y}^0$ must be optimal. If $h > 0$, there will be a lowest integer $j^* \notin I$ such that equality is attained in (7), and since 1 and n can never be deleted from I, it must be the case that $2 \le j^* \le n - 1$. Then

$$y^0_{j^*} = f_{j^*} - h_0,$$

and $k = j^-(I)$ and $k^+ = j^+(I)$ are consecutive elements of I such that $k < j < k^+$ and

$$y^0_k = f_k + h_0,$$

$$y^0_{k^+} = f_{k^+} + h_0.$$

Since $j^\star \notin I$, $c_{k,j^\star,k^+}(\mathbf{y}^0) = c_{k,j^\star,k^+}(\mathbf{f}(I)) = 0$. Now if $\mathbf{v} \in \mathbb{R}^n$ and $F(\mathbf{v}) < h$, then

$$v_k < f_k + h = y_k^0,$$

$$v_{j^\star} > f_{j^\star} - h = y_{j^\star}^0,$$

$$v_{k^+} < f_{k^+} + h = y_{k^+}^0.$$

It follows that $c_{k,j^\star,k^+}(\mathbf{v}) < c_{k,j^\star,k^+}(\mathbf{y}^0) = 0$. It now follows from Theorem 1 that $\mathbf{v} \notin Y_0$. Therefore $h = h_0$ and $\mathbf{y}^0$ is optimal.   □

The concave case follows easily.

Corollary 2 *The components of a best* concave *approximation to data* f_j, $r \le j \le s$, *are given from Algorithm 1 with* $\sigma = -$ *by* $y_j = f_j(I) - h$, *where in this case* $I = I^-(r, s)$ *and* $h = h^-(r, s)$.

The algorithms in the next subsections will join optimal vertex sets produced by Algorithm 1 of consecutive ranges of the data, and they will also split such optimal vertex sets in two. It is convenient to prove here that the resulting sets remain optimal vertex sets. The proof requires one further important property of the optimal vertex sets produced by Ubhaya's algorithm.

Let k be the second element of $I^+(1, n)$. It is easy to show from the definition of the lower convex hull, as embodied in (10) and (11), that the gradient g_{1k} is minimal, namely that $g_{1k} = \min\{g_{1j} : 2 \le j \le n\}$. This principle can be applied recursively to all the elements of $I^+(1, n)$. It also applies backwards starting with the penultimate element of $I^+(1, n)$. Applying this to a general range of the data produces the result (which will be much used later) that if i and k are consecutive elements of $I^\sigma(r, s)$ then

$$\sigma g_{ik}(\mathbf{f}(I)) = \min\{\sigma g_{ij}(\mathbf{f}(I)) : i \le j \le s\} \qquad \text{and} \qquad (12)$$

$$\sigma g_{ik}(\mathbf{f}(I)) = \min\{\sigma g_{jk}(\mathbf{f}(I)) : r \le j \le i\}. \qquad (13)$$

This minimality principle will be used immediately below. It can be seen as defining the optimal vertex sets recursively and could be applied to construct them, but such a method would not be as efficient as the systematic deletion algorithm of Ubhaya.

The results for the joining and splitting of optimal vertex sets require the definition of trivial optimal vertex sets by

$$I^\sigma(r, s) = [r, s], \qquad \text{for } r \ge s - 1. \qquad (14)$$

It is also convenient to define $h^\sigma(r, s) = 0$, for $r \ge s - 1$. The following lemmas then hold.

Lemma 1 *A subset of one or more consecutive elements of an optimal vertex set is itself an optimal vertex set.*

Proof The proof is trivial when the subset has fewer than three elements. Otherwise it follows from the nature of extreme points of convex sets. The elements are recursively specified by (12) and (13) and the minima are the same when taken over a more restricted range. For example, if $I^+(1, n) = \{1, k, s, \ldots, n\}$, then g_{1k} minimizes g_{1j} over the range $2 \leq j \leq n$ and so also over the range $2 \leq j \leq s$. □

The second lemma gives conditions for the amalgamation of optimal vertex sets.

Lemma 2 *Given vertex sets $I^\sigma(r, s)$ and $I^\sigma(s, t)$ necessary and sufficient conditions for*

$$I^\sigma(r, t) = I^\sigma(r, s) \cup I^\sigma(s, t) \tag{15}$$

are that there exist $r' \leq r$, $r' \in I^\sigma(r, s)$, and $t' \geq t$, $t' \in I^\sigma(s, t)$, such that

$$s \in I^\sigma(r', t'). \tag{16}$$

Proof Necessity follows directly in all cases by letting $r' = r$ and $t' = t$, immediately giving (16). Sufficiency is trivial except when $r < s < t$. Let s^- be the left neighbour of s in $I^\sigma(r, s)$ and s^+ its right neighbour in $I^\sigma(s, t)$. Then $r' \leq s^-$ and $t' \geq s^+$. It follows from the minima in (12) and (13) that (15) holds provided that

$$\sigma c_{s^-, s, s^+}(\mathbf{f}) > 0.$$

However, it follows from (16) that $\sigma c_{isj}(\mathbf{f}) > 0$ for *all* i and j in the range $r' \leq i < s < j \leq t'$, and this range contains s^- and s^+. □

All the results required for the next subsections have now been established. There is, however, an interesting theoretical consequence of the minimality principle expressed in (12) and (13). It turns out that there is a remarkable equivalence between the problem of this subsection—an ℓ_∞ minimization subject to convexity constraints—and quite a different constrained optimization problem. The convexity constraints are equivalent to the gradients of the smoothed data being monotonically increasing and it turns out that the optimal vertex set constructed above is also defined by the solution of a weighted ℓ_2 optimization problem subject to monotonicity constraints. Consider the problem of minimizing the function

$$G(z) = \sum_{j=1}^{n-1} w_j(z_j - g_j)^2,$$

for given values g_j and non-negative weights w_j, subject to the constraints $z_1 \leq \cdots \leq z_{n-1}$. Such problems were considered earlier in Cullinan [1]. When the material for this paper was being produced, it was recalled that Ubhaya's algorithm was a specialization of an algorithm by Graham [6] for finding the convex hull of an unordered set of points in the plane. Ubhaya's algorithm has the same logical structure as the algorithm of Kruskal [8] for monotonic rather than convex approximation. Kruskal's algorithm is more efficient than the algorithm of Miles [9] for monotonic approximation, but produces the same results. Graham's algorithm constructs the solution through blocks of equal values. The value in each block is the best ℓ_2 approximation by a constant to the data in the range of that block, i.e. a weighted mean of the data in that block. The solution can therefore be expressed in terms of a set of indices at which the constraints are unequal. If i and k are two consecutive unequal constraint indices, or end indices, then the z_j, $i \leq j \leq k$ are given by $z_j = a_{ik}$ where

$$a_{ik} = \frac{\sum_{j=i}^k w_j g_j}{\sum_{j=i}^k w_j}.$$

Now let the g_j be defined by $g_j = g_{j\,j+1}(\mathbf{f})$ and the weights by $w_j = x_{j+1} - x_j$. A short calculation shows that $a_{ik} = g_{ik}$, and so Graham's algorithm will actually calculate the optimal vertex set for the main problem of this subsection. The proof of this interesting equivalence is a straightforward consequence of the logically equivalent structures of Kruskal's and Miles's algorithms for monotonic approximation. It is given in Cullinan [1]. Thus you can actually do an ℓ_∞ optimization subject to convexity by doing a weighted ℓ_2 monotonic optimization on the gradients of the data!

The convex–concave case can now be discussed.

2.2   The Case $q = 1$

The algorithm to be presented constructs data ranges $[1, s]$ and $[t, n]$, where $s \leq t$, a price $h \geq 0$, and a vertex set I of $[1, n]$ such that

$$I = I^+(1, s) \cup I^-(t, n).$$

The best approximation $\mathbf{y}^1$ is then given as the final value of the vector $\mathbf{y}$ defined from these quantities by

$$y_i = f_i + h \qquad \text{when } i \in I^+(1, s) \tag{17}$$

$$y_i = f_i - h \qquad \text{when } i \in I^-(t, n) \tag{18}$$

$$y_j = y_j(I) \qquad 1 \leq j \leq n. \tag{19}$$

It will be shown that $s = t$ only if $h = 0$, so that $\mathbf{y}^1$ is well defined. This construction is illustrated in Figure 2. Note that when $s = t = n$, $\mathbf{y} = \mathbf{y}^0$.

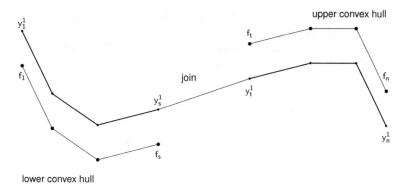

Fig. 2 Construction of best convex–concave approximation

The algorithm to be given is believed to be more efficient than that in Cullinan and Powell [3]. For example, if the data are in Y_0, the new algorithm will require only one iteration, whereas the former algorithm only has this property if the data lie on a straight line.

The algorithm builds up I by looking alternately at the left and right ranges and joining segments of the convex hulls when it is best to do so. Beginning with $I = \{1, n\}$, $s = 1$, and $t = n$, it adds an index k of $I^+(s, t)$ to I if the least possible final value of F consistent with doing this is not greater than the least possible value of F consistent with ending the calculation with the existing value of s. After adding one index in $I^+(s, t)$ to I and increasing s to the value of this index, it then examines the next. When it is not worth adding any more indices from $I^+(s, t)$ to I, it tries to add indices in $I^-(s, t)$ to I working backwards from t, adding k to I if it is not necessarily more expensive to finish with t reduced to k than with t as it is and decreasing t. After indices have been added to I from $I^-(s, t)$ it may then be possible to add more to I from the new $I^+(s, t)$, so the process alternates between $I^+(s, t)$ and $I^-(s, t)$ until s equals t or $t - 1$ or until the algorithm fails twice running to add any indices.

Algorithm 2 *To find the optimal vertex set, ranges, and price of a best convex–concave approximation.*

Step 1. Set $s := 1$, $t := n$, $h := 0$, and $I := \{1, n\}$.

Step 2. If $s \geq t - 1$: **stop.** Otherwise: set $u = t - s$ and use Algorithm 1 to calculate $I' := I^+(s, t)$.

Step 3. Let $s^+ = s^+(I')$. Calculate $h^+(s, s^+)$ from (7) and set $h' := \max(h, h^+(s, s^+))$.
Calculate $f_t(s, s^+)$ from (6). If $f_t(s, s^+) > f_t - 2h'$: go to Step 5.

Step 4. Add s^+ to I and delete s from I'. Set $h := h'$ and $s := s^+$.
         If $s < t$: go back to Step 3.
Step 5. If $s \geq t - 1$: **stop.** Otherwise: use Corollary 2 to calculate
         $I^-(s,t)$ and set $I' := I^-(s,t)$.
Step 6. Let $t^- = t^-(I')$. Calculate $h^-(t^-, t)$ from (7) and set $h' :=$
         $\max(h, h^-(t^-, t))$.
         Calculate $f_s(t^-, t)$ from (6). If $f_s(t^-, t) < f_s + 2h'$: go to
         Step 8.
Step 7. Add t^- to I and delete t from I'. Set $h := h'$ and $t := t^-$.
         If $s < t$: go back to Step 6.
Step 8. If $t - s = u$: **stop.** Otherwise: go back to Step 2.

The complexity of this algorithm is $n \log n$. The proof of its effectiveness will follow the same course as that of the last theorem. Thus the first remark is that s is non-decreasing, t is non-increasing, and $s \leq t$ with $s = t$ only when $h = 0$. The vector $\mathbf{y}^1$ is therefore always well defined by (17)–(19).

The possibility that Algorithm 2 can end with $s = t$ will create slightly more complexity when this algorithm is used later on when $q \geq 3$. It might be prevented by relaxing either of the inequalities in Steps 3 and 6. However, if *both* these inequalities are relaxed, then the algorithm will fail. For example, with data at equally spaced abscissæ and $\mathbf{f} = (0, 1, 0, 1)$, relaxing both inequalities yields $s = 1$, $t = 4$, and $F(\mathbf{y}) = \frac{2}{3}$, whereas the Algorithm as it stands correctly calculates $h = \frac{1}{2}$ and $\mathbf{y} = (\frac{1}{2}, \frac{1}{2}, \frac{1}{2}, 1)$. It seemed best to let both inequalities be strict for reasons of symmetry. This feature does, however, raise the possibility that when the data contain a parallelogram rounding errors may cause the algorithm to fail.

The next result is a lemma concerning the conditions under which Algorithm 2 increases s and decreases t. It will be used to show that the final value of h is not determined by data in the join. Its somewhat cumbersome statement is needed to include the case where $h = 0$.

Lemma 3 *At any entry to Step 2 of Algorithm 2, let the vector $\mathbf{y}$ be defined by (17)–(19). Then the algorithm strictly decreases $t - s$ if and only if*

$$\text{there exists } j : s < j < t \quad \text{and} \quad |f_j - y_j| \geq h. \tag{20}$$

Since the points (x_j, y_j), $s \leq j \leq t$, are collinear, (20) is equivalent to the statement that there exists a point of the graph of the data between x_s and x_t lying on or outside the parallelogram $\Pi(h)$ with vertices (x_s, f_s), $(x_s, f_s + 2h)$, (x_t, f_t), and $(x_t, f_t - 2h)$. This parallelogram $\Pi(h)$ is illustrated in Figure 3.

Proof In the trivial case when $s \geq t - 1$, (20) is false and Algorithm 2 stops without altering s or t.

Otherwise, there are one or more data points f_j with $s < j < t$. Suppose first that (20) does not hold. Then $h > 0$. (If $h = 0$, then (20) holds trivially!) It will be shown that in this case both s and t are left unchanged. Step 3 will calculate an index $s^+ : s < s^+ \leq t$.

Fig. 3 The parallelogram
$\Pi(h)$

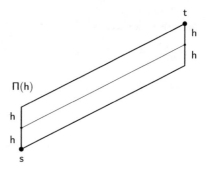

If $s^+ = t$, then because $h' \geq h > 0$, it immediately follows that $f_t(s, s^+) = f_t > f_t - 2h'$, so that Step 3 will lead immediately to Step 5 and s will not be increased.

If $s^+ < t$, let y_{s^+} be defined by (17)–(19) and let $g = g_{st}(\mathbf{y})$. Then by hypothesis, $f_{s^+} > y_{s^+} - h$, which implies that $g_{s\ s^+}(\mathbf{f}) > g$. Thus,

$$f_t(s, s^+) = f_s + g_{s\ s^+}(\mathbf{f})(x_t - x_s)$$
$$> f_s + g(x_t - x_s) = y_t - h = f_t - 2h$$
$$\geq f_t - 2h'.$$

Therefore Step 3 will again lead immediately to Step 5.

The same arguments applied to t^- calculated by Step 6 show that Step 6 will branch immediately to Step 8 and so t will also be left unchanged.

Now suppose conversely that (20) does hold, so that there is an index j with data point lying on or outside $\Pi(h)$. It will be shown that in this case if the algorithm does not increase s, it must then decrease t.

It suffices to consider in detail only the case where $f_j \leq y_j - h$ (which is illustrated in Figure 4) because the case where $f_j \geq y_j + h$ is exactly similar. Step 3 will calculate an index s^+ in the range $s < s^+ \leq t$, and, in view of (12), with $g_{s\ s^+}(\mathbf{f}) \leq g_{sj}(\mathbf{f})$. Then

$$f_t(s, s^+) = f_s + g_{s\ s^+}(\mathbf{f})(x_t - x_s)$$
$$\leq f_s + g_{sj}(\mathbf{f})(x_t - x_s)$$
$$= f_s + ((f_j - f_s)/(x_j - x_s))(x_t - x_s)$$
$$\leq f_s + ((y_j - h - f_s)/(x_j - x_s))(x_t - x_s) \text{ by hypothesis}$$
$$= y_s - h + ((y_j - y_s)/(x_j - x_s))(x_t - x_s)$$
$$= y_t - h, \text{ by the definition of } y_j.$$

Fig. 4 When s does not
increase, t must decrease

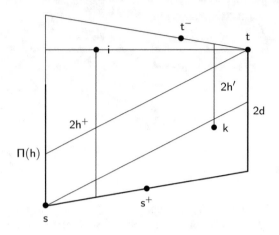

There are two alternatives to consider depending on whether $h' = h$ or $h' > h$. In
the former case when $h' = h$ it now follows immediately that Step 4 will be entered
and s increased to s^+, as required.

Now suppose that $h' = h^+(s, s^+) > h$ and that s is not increased, i.e. that the
test in Step 3 leads to Step 5. Then $f_t(s, s^+) > f_t - 2h' = f_t - 2h^+(s, s^+)$. Let
$h^+ = h^+(s, s^+)$. By definition of h^+, there must be an index i such that $f_i -
f_i(s, s^+) = 2h^+$. This case as illustrated in Figure 4 demonstrates the heart of the
principle behind the algorithm and the essence of this lemma because the four data
points with indices s, i, s^+, and t determine a lower bound on $\inf F(Y_1)$ and this
lower bound is never determined by data left in the join between the final values of
f_s and f_t. Define

$$2d = f_t - f_t(s, s^+). \tag{21}$$

Then by the assumption that Step 3 led to Step 5,

$$h^+ > d. \tag{22}$$

Step 5 will be entered with $h < h^+$ and will calculate $I^-(s, t)$. Step 6 will then find
an index t^- such that $s \le t^- < t$. It follows from (12) that $f_i \le f_i(t^-, t)$, and so

$$2h^+ = f_i - f_i(s, s^+) \le f_i(t^-, t) - f_i(s, s^+).$$

This inequality along with (21) and (22) then shows that the function $l \mapsto
f_l(t^-, t) - f_l(s, s^+)$ is strictly decreasing. Therefore in particular, $f_s(t^-, t) -
f_s(s, s^+) > 2h^+$.

If h does not increase in Step 7, it follows immediately, from the assumption that
$h < h^+$ and the last inequality, that $f_s(t^-, t) - f_s = f_s(t^-, t) - f_s(s, s^+) > 2h^+ >
2h$, so that t will be reduced to t^- by the test in Step 6.

If, on the other hand, h does increase in Step 7, its new value h' is given by $2h' = f_k(t^-, t) - f_k$ for some k in the range $s < k < t$. Since the definition of s^+ implies that $f_k \geq f_k(s, s^+)$ it follows that $2h' \leq f_k(t^-, t) - f_k(s, s^+) < f_s(t^-, t) - f_s(s, s^+)$, so that again t must be reduced to t^- by Step 7. In fact it is easy to show that in this case, $t^- \geq i$, $h' < h^+$, and that inf $F(Y_1) \geq \frac{1}{2}(f_{s^+}(i, t) - f_{s^+})$.

Thus t will be reduced to t^- in all cases.

In the case where there exists j such that $f_j \geq y_j + h$, if s is not increased immediately after the next entry to Step 3, the same argument shows that either t must be reduced in the next operation of Steps 6 and 7 or s must be increased immediately thereafter.   □

The next lemma is needed to establish the feasibility of $\mathbf{y}^1$ by establishing that the join constraints do not bend the wrong way.

Lemma 4 *At any exit from Step 4 or Step 7, let $\mathbf{y}$ be defined by (17)–(19). Then*

$$\text{when } s > 1, \quad c_{s-1}(\mathbf{y}) \geq 0, \tag{23}$$

and

$$\text{when } t < n, \quad c_{t-1}(\mathbf{y}) \leq 0. \tag{24}$$

Proof The proof will be by induction. There is nothing to prove unless Step 3 is entered at least once. Suppose first that Step 3 is entered and leads to Step 4. Then (23) at exit from Step 4 is equivalent to

$$c_{s^+-1}(\mathbf{y}) \geq 0,$$

which is in turn equivalent to the identity

$$f_t - f_t(s, s^+) \geq 2h'. \tag{25}$$

But this is simply the test leading from Step 3 to Step 4.

Repeated application of this argument shows that whenever Step 5 first branches to Step 6 then (23) remains true. This result will be useful in the cases $q > 1$.

The next stage is to establish (24) when Step 3 leads to Step 4. If $t < n$, there will exist an index $t^+ = t^+(1)$, and (24) will be equivalent to the inequality

$$f_{s^+}(t, t^+) - f_{s^+} \geq 2h'. \tag{26}$$

Now define the monotonic function $\phi : l \mapsto f_l(t, t^+) - f_l(s, s^+)$. Then (26) is equivalent to $\phi(s^+) \geq 2h'$.

Suppose firstly that $h' > h$, so that there exists i such that $s < i < s^+$ and

$$f_i - f_i(s, s^+) = 2h'. \tag{27}$$

The earlier definition of t from t^+ as an index in the set specifying the upper convex hull of data to the left of t^+ implies that $f_i \leq f_i(t, t^+)$, so it follows from (27) and (25) that the monotonic function ϕ satisfies $\phi(i) \geq 2h'$ and $\phi(t) \geq 2h'$. Therefore $\phi(s^+) \geq 2h'$ as required to establish (26).

The inductive assumption was not needed in this case. It follows that whenever h increases, *both* the join constraints c_{s-1} and c_{t-1} (when defined) point the right way for feasibility. This point will also be important in the general case when $q > 1$.

When Step 4 is entered and h does not increase, it is necessary to assume inductively that c_{t-1} is non-positive initially, before s is increased. This is equivalent to the inequality

$$f_s(t, t^+) - f_s \geq 2h \tag{28}$$

or $\phi(s^+) \geq 2h$. It follows from this and (25) with $h' = h$ that $\phi(t) \geq 2h$, so that again $\phi(s^+) \geq 2h$ as required.

The same argument shows that when Step 7 is entered and h does not increase, if c_{s-1} is initially non-negative, it will remain so when s is increased.

The result then follows by induction.   □

When $q > 1$ the algorithms proposed below will carry out the procedure of Steps 2 to 8 of Algorithm 2 but possibly starting with a positive value of h. The proof of Lemma 4 shows that once h is increased feasibility continues to hold, thus establishing the following corollary.

Corollary 3 *If steps 2 to 8 of Algorithm 2 are executed with h initially set to any positive number, then (23) and (24) remain satisfied.*

The effectiveness of Algorithm 2 can now be established.

Theorem 3 *Algorithm 2 produces integers s and t with $s \leq t$, a vertex set I such that*

$$I = I^+(1, s) \cup I^-(t, n), \tag{29}$$

and a real number h such that $h = 0$ if and only if $s = t$. If $\mathbf{y}^1$ is then defined by (17)–(19), then $\mathbf{y}^1 \in Y_1$ and $F(\mathbf{y}^1) = h = \inf F(Y_1)$.

Proof The proof proceeds as in the case of Theorem 2. First it is shown that a well-defined vector $\mathbf{y}^1$ is produced, then that this vector is feasible, and finally that it is optimal because no vector with a lower value of the objective function F can be feasible.

The first stage in showing that $\mathbf{y}^1$ is well defined is to establish (29). Assume inductively that it holds before a series of sections is added, and without loss of generality that a series of convex sections with indices $\{s, s_1, \ldots, s_\alpha\}$ are added by successive entries to Step 4 for the same value of t. Each time an index is deleted from I' it follows immediately from Lemma 1 that the new value of I' is also an optimal vertex set, so it always holds that $I' = I^+(s, t)$. It also follows from Lemma 1 that $\{s, s_1, \ldots, s_\alpha\} = I^+(s, s_\alpha)$. Let $I_1 = \{i \in I : i \leq s\}$. After all the sections are added, $I_1 = I^+(1, s) \cup I^+(s, s_\alpha)$. If $s = 1$, it follows at once from Lemma 2 that $I_1 = I^+(1, s_\alpha)$ as required. Otherwise, let t' be the value that t had when s was increased from s^-. Then $s_\alpha \leq t \leq t'$. Immediately after s was increased from s^-, $s \in I^+(s^-, t')$. The conditions of Lemma 2 are therefore satisfied and $I_1 = I^+(1, s_\alpha)$. The same argument applied to concave sections then establishes (29).

Algorithm 2 clearly produces a number h such that $s = t$ only if $h = 0$. It is a consequence of Lemma 3 that if $s < t - 1$ and $h = 0$, then the algorithm will reduce $t - s$. If $s = t - 1$ and $h = 0$, Step 4 will increase s to t. Thus $h = 0$ if and only if $s = t$. Thus $\mathbf{y}^1$ is well defined by (17)–(19).

The number h is given by $h = \max(h_{(1)}, h_{(2)})$ where

$$h_{(1)} = h^+(1, s),$$

$$h_{(2)} = h^-(t, n).$$

It follows from (29) and Lemma 3 that when the algorithm terminates,

$$F(\mathbf{y}^1) = h. \tag{30}$$

It then follows directly from (29), (17)–(19) and Lemma 4 that $\mathbf{y}^1 \in Y_1$.

It remains to prove that $\mathbf{y}^1$ is optimal. The method of proof chosen to do this can be simplified in this case, but generalizes more directly to the case of $q > 1$. If $h = 0$, optimality follows immediately from (30). Otherwise suppose that there is a vector $\mathbf{v}$ such that $F(\mathbf{v}) < h$. The price h will be defined from (7) with particular values of σ, r, and s. Let j^* be the lowest value of j in this equation that defines the final value of h. Then j^* lies strictly between two neighbouring elements k and k^+ of I. Assume firstly that $j^* < s$. Define the set $K = \{k, j^*, k^+, s, t\}$. Since $h > 0$, $s < t$, so K has at least four elements (it is possible that $k^+ = s$). Now

$$y_k^1 = f_k + h,$$
$$y_{j^*}^1 = f_{j^*} - h,$$
$$y_{k^+}^1 = f_{k^+} + h,$$
$$y_s^1 = f_s + h,$$
$$y_t^1 = f_t - h.$$

Since $F(\mathbf{v}) < h$,

$$v_k < f_k + h,$$
$$v_{j^*} > f_{j^*} - h,$$
$$v_{k^+} < f_{k^+} + h,$$

$$v_s < f_s + h,$$
$$v_t > f_t - h.$$

By definition of $y^1_{j^\star}$,

$$c_{kj^\star k^+}(\mathbf{y^1}) = 0,$$

while from the above inequalities $c_{kj^\star k^+}(\mathbf{v}) < c_{kj^\star k^+}(\mathbf{y^1}) = 0$. From Lemma 4, $c_{i-1}(\mathbf{y^1}) \geq 0$ for all i in the range $j^\star \leq i \leq t-1$. It follows from the corollary to Theorem 1 that

$$c_{j^\star st}(\mathbf{y^1}) \geq 0.$$

Then $c_{j^\star st}(\mathbf{v}) > c_{j^\star st}(\mathbf{y^1}) \geq 0$. If $k^+ = s$, Theorem 1 can be immediately applied to K to show that $\mathbf{v} \notin Y_1$. If $k^+ < s$, it follows from the corollary to Theorem 1 that because $c_{j^\star st}(\mathbf{v}) > 0$, at least one of the consecutive divided differences $c_{j^\star k+s}(\mathbf{v})$ and $c_{k+st}(\mathbf{v})$ must be positive, so that again $\mathbf{v} \notin Y_1$.

If $j^\star > s$, let $K = \{s, t, k, j^\star, k^+\}$. Then the same argument shows that $\mathbf{v}$ cannot be feasible. Therefore $\mathbf{y}^1$ is optimal. $\square$

It is clear that Algorithm 2 can easily be altered to calculate a best concave-convex approximation instead of a convex–concave one.

It is also worth noting here that a global minimum solution to the $q = 1$ optimization problem has been constructed.

2.3   The Case $q = 2$

The best Y_1 approximation constructed in Section 2.2 was defined by Equations (17)–(19) from the two pieces of the data in the ranges $[1, s]$ and $[t, n]$, the vertex set I, and the price h. The best Y_2 approximation will in general be constructed from three pieces of the data in the ranges $P_1 = [1, s_1]$, $P_2 = [t_1, s_2]$, $P_3 = [t_2, n]$, a price $h \geq 0$ with $h > 0$ only when $s_1 < t_1$ and $s_2 < t_2$, and a vertex set I of $[1, n]$ such that

$$I = I^+(P_1) \cup I^-(P_2) \cup I^+(P_3), \tag{31}$$

as the ultimate value $\mathbf{y}^2$ of the vector $\mathbf{y}$ defined by the equations

$$y_i = f_i + (-)^{\alpha-1}h \qquad \text{when } i \in I \cap P_\alpha, \qquad 1 \leq \alpha \leq 3, \tag{32}$$
$$y_j = y_j(I) \qquad 1 \leq j \leq n. \tag{33}$$

The construction of $\mathbf{y}$ is illustrated in Figure 5.

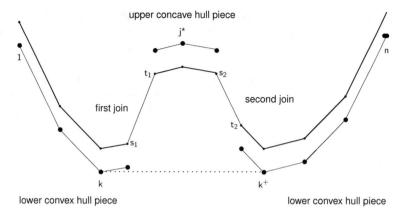

Fig. 5 Construction of Y_2 approximation

The Algorithm will construct this best Y_2 approximation from the quantities $h_0 = h^+(1, n)$ and the index set $I^+(1, n)$ provided by Algorithm 1. If the value h_0 of this approximation is zero, the best approximation over Y_0 is also a best approximation over Y_2. Otherwise, $h_0 > 0$ is determined by three data f_k, f_{j^*}, and f_{k^+} such that k and k^+ are consecutive elements of $I^+(1, n)$ and $k < j^* < k^+$. The discussion in Section 2.1 shows that unless the divided differences of $\mathbf{y}$ change sign at least once in the range $[x_k, x_{k^+}]$, then $F(\mathbf{y}) \geq h_0$. The set $I^-(t_1, s_2)$ is therefore put in this range. The algorithm begins with $s_1 = k$, $t_1 = s_2 = j^*$, and $t_2 = k^+$. It then sets $h = \max(h^+(1, s_1), h^+(t_2, n))$. Next, it uses Algorithm 2 modified to calculate a best convex–concave approximation to the data with indices in the range $[1, j^*]$ consistent with paying this minimum price h. It is an important feature of the problem that this can be done by starting Algorithm 2 with $s = k$ and $t = j^*$, i.e. the existing elements of $I^+(1, j^*)$ below k can be kept in place.

This process can increase s_1 beyond k, reduce t_1 below j^*, and increase h. Let $h^{(1)}$ be its new value. A best *concave-convex* approximation to $f_{j^*}, \ldots, f_n$ starting with $h = h^{(1)}$ is next identified by applying a modified version of Algorithm 2 with $s = j^*$ and $t = k^+$, in general increasing s_2 beyond j^* and reducing t_2 below k^+. If this second calculation does not increase h above $h^{(1)}$, the best approximation $\mathbf{y}$ can be constructed immediately from (32)–(33). If, however, $h > h^{(1)}$, there is the complication that the lower value of h when the first calculation took place may have joined too many sections for the first join constraints $c_{s_1-1}(\mathbf{y})$ and $c_{t_1-1}(\mathbf{y})$ determined by the new higher value of h to have the right signs. In this case the remedy proposed is to repeat the first calculation starting with the new value of h.

The following algorithms will therefore require a modified version of Algorithm 2 to carry out a best convex–concave approximation or a best concave-convex approximation on a range of data, starting with a prescribed value of h. This task is best carried out by modifying Algorithm 2 in detail, but the following description is equivalent and simpler. The following procedure calculates a best approximation to the data in the range $[s_\alpha, t_\alpha]$ compatible with an existing price h, the approximation being over Y_1 or Y_{-1} according as α is odd or even. It can be seen as trying to close the join range $[s_\alpha, t_\alpha]$ as much as possible by constructing the best approximation to this range of data compatible with the given starting value of h.

The notation for the pieces of the data is based on the observation that when $h > 0$, $t_\alpha = s_\alpha^+(I)$ and when $h = 0$, $t_\alpha = s_\alpha$, so that the location of each of the two join ranges $[s_1, t_1]$ and $[s_2, t_2]$ can be specified simply by consecutive elements of I. Thus the location of the pieces and joins can be specified simply through the quantities s_α, $1 \le \alpha \le 2$. It is convenient to regard these as members of an ordered subset S of I, and to include n in S. Given a vertex set I, define a *piece set* S of I to be an ordered subset of I such that $n \in S$. Then s_α will denote the αth element of S.

Algorithm 3 *closejoin(α): modifying h, I, and S.*

Step 1.   If α is even: replace **f** by $-$**f**.
            Set $s = s_\alpha$ and $t = s_\alpha^+(I)$.
Step 2.   Carry out Steps 2 to 8 of Algorithm 2.
Step 3.   Set $s_\alpha = s$.
            If α is even: replace **f** by $-$**f**.

The following algorithm constructs $\mathbf{y}^2$ by calculating the appropriate pieces and a price $h(S)$ from which $\mathbf{y}^2$ is constructed. The notation used for keeping track of the pieces needs to cover the case of pieces that consist of only one point, and to generalize easily when $q > 2$. It also has to cope with the trivial cases where $\mathbf{f} \in Y_0$ or $\mathbf{f} \in Y_1$ and there are, therefore, only one or two pieces instead of three.

Given a piece set S, define its price $h(S)$ as follows. Let $q' = |S| - 1$, $t_0 = 1$, and when $q' \ge 1$ define $t_\alpha = s_\alpha^+(I)$ when $1 \le \alpha \le q'$. Then

$$h(S) = \max\{ h^{(-)^{\alpha-1}}(t_{\alpha-1}, s_\alpha) : 1 \le \alpha \le q' + 1 \}. \qquad (34)$$

Recall that $h^\sigma(t, s) = 0$ whenever $t \ge s - 1$.

It is also worth recording the location of the data points determining the optimal value of h. Given a piece set S, when $h(S) > 0$ it follows from (34) and (7) that there exists a lowest index $j^\star(S)$ such that

$$h(S) = \tfrac{1}{2}\sigma(f_{j^\star} - f_{j^\star}(k, k^+)), \qquad (35)$$

where $\sigma = (-)^{\beta-1}$ for some β in the range $1 \le \beta \le q' + 1$, and

$$t_{\beta-1} \le k < j^\star < k^+ \le s_\beta. \qquad (36)$$

Once $j^\star$ is known, the quantities k, k^+, and β are uniquely determined by (35) and (36).

The following algorithm then constructs h, I and S determining $\mathbf{y}^2$.

Algorithm 4 *To find the vertex set, price, and piece set of a best approximation over Y_2.*

Step 1.   Set $S = \{n\}$. Use Algorithm 1 to calculate $I = I^+(1, n)$ and
$h = h^+(1, n)$.
If $h = 0$: **stop.**
Otherwise: find $j^\star, k, k^+$ determined from S by (35) and proceed to Step 2.

Step 2.   Insert $j^\star$ into I and $k, j^\star$ into S. Calculate $h = h(S)$.

Step 3.   Apply *closejoin*(1). Set $h^{(1)} = h$.

Step 4.   Apply *closejoin*(2). If $h = h^{(1)}$: **stop.**
Otherwise proceed to Step 5.

Step 5.   If $h^{(1)} = 0$: go to Step 6.
If $s_1 = k$ and $s_1^+(I) = j^\star$: **stop.**
Otherwise set $s = s_1$ and $t = s_1^+$ and calculate $g^{(1)} = (f_t -$
$f_s - 2h)/(x_t - x_s)$.
If $s > k$ and $g_{s^-(I)s} > g^{(1)}$: go to Step 6.
If $t < j^\star$ and $g_{t\,t^+(I)} > g^{(1)}$: go to Step 6.
Otherwise: **stop.**

Step 6.   Set $s_1 = k$. Delete all elements of I lying strictly between k and
$j^\star$ and then apply *closejoin*(1).

It will be seen that, as described above, Step 2 begins the new approximation, Steps 3 and 4 carry out the convex–concave and concave-convex approximations, Step 5 tests whether feasibility has been violated and, if it has, Step 6 repeats the convex–concave approximation with the new value of h. Since Algorithm 4 calls Algorithm 1 once and *closejoin* up to three times, its complexity is that of *closejoin*, which is the same as Algorithm 2, namely $n \log n$.

Now define the vector $\mathbf{y}(S)$ from S and h as follows. Let $q' = |S| - 1$ and $t_0 = 1$. When $q' \geq 1$ define $t_\alpha = s_\alpha^+(I)$ when $h > 0$ and $t_\alpha = s_\alpha$ when $h = 0$, for $1 \leq \alpha \leq q'$. Then let

$$P_\alpha = [t_{\alpha-1}, s_\alpha], \qquad \text{when } 1 \leq \alpha \leq q' + 1, \qquad (37)$$

and define $\mathbf{y}(S)$ by (32)–(33). Set $\mathbf{y}^2$ to the value of $\mathbf{y}(S)$ on exit from Algorithm 4.

Step 5 is designed to avoid calculating gradients unnecessarily. The following lemma will be used to justify this economy and also to show that $\mathbf{y}^2$ is feasible. Recall the parallelogram $\Pi(h)$ defined in the proof of Lemma 3, and given I and S, let $s = s_\alpha$ and $t = s_\alpha^+(I)$ and define $\Pi_\alpha^S(h)$ as the closed solid parallelogram with vertices (x_s, f_s), $(x_s, f_s + 2(-1)^{\alpha-1}h)$, (x_t, f_t), and $(x_t, f_t - 2(-1)^{\alpha-1}h)$. Each such parallelogram then defines a join gradient

$$g^{(\alpha)}(S, h) = \frac{f_t - f_s - 2(-1)^{\alpha-1}h}{x_t - x_s}. \qquad (38)$$

Lemma 5 *Let closejoin* (α) *be called, modifying* h', I', *and* S' *to* h, I, *and* S, *respectively. Suppose that there exists* $\overline{h} \geq 0$ *such that*

$$(x_j, f_j) \in \Pi_\alpha^{S'}(\overline{h}) \qquad for \quad s'_\alpha \leq j \leq s'^+_\alpha(I'). \tag{39}$$

and that

$$h' \leq \overline{h}. \tag{40}$$

Then

$$h \leq \overline{h}. \tag{41}$$

Further, if $\overline{g} = g^{(\alpha)}(S', \overline{h})$, $g^+ = g_{s'_\alpha}(\mathbf{y}(S))$ *and* $g^- = g_{s'^+_\alpha(I')-1}(\mathbf{y}(S))$, *then*

$$(-1)^{\alpha-1} \min(g^+, g^-) \geq (-1)^{\alpha-1}\overline{g}. \tag{42}$$

Proof Note that g^+ and g^- are the new gradients at the *old* join points, for example, k and $j^\star$.

Assume for simplicity that α is odd and write $s' = s'_\alpha$ and $t' = s'^+_\alpha(I')$.

First consider (41). The proof is trivial unless *closejoin* increases h. In this case, $h > 0$ and so there exist $j^\star$, k, $k^+ \in I$ such that

$$2h = f_{j^\star} - f_{j^\star}(k, k^+), \qquad \bullet \tag{43}$$

where $s' \leq k < j^\star < k^+ \leq t'$. Let $t_\alpha = s'^+_\alpha(I)$. It follows from Lemma 3 that either $j^\star \leq s_\alpha$ or $j^\star \geq t_\alpha$. The two cases are entirely similar: it suffices to consider the first. Define $\overline{y}_j$, $s' \leq j \leq t'$, by $\overline{y}_{s'} = f_{s'} + \overline{h}$, $\overline{y}_{t'} = f_{t'} - \overline{h}$, and $\overline{y}_j = \overline{y}_j(s', t')$, $s' < j < t'$. Then the $\overline{y}_j$, $s' \leq j \leq t'$, are collinear and it follows from (39) that $f_k \geq \overline{y}_k - \overline{h}$ and $f_{k^+} \geq \overline{y}_{k^+} - \overline{h}$. Therefore, since $\overline{y}_k$, $\overline{y}_{j^\star}$, and $\overline{y}_{k^+}$ are collinear,

$$\overline{y}_{j^\star} - \overline{h} \leq f_{j^\star}(k, k^+). \tag{44}$$

It also follows from (39) that

$$f_{j^\star} \leq \overline{y}_{j^\star} + \overline{h}. \tag{45}$$

Addition of (44) and (45) and use of (43) then gives (41).

Now consider (42). The easiest case is when $s' < s_\alpha$. Then $g^+ = g_{s'i}$, where $i = s'^+(I)$. Let the $\overline{y}_j$, $s' \leq j \leq t'$, be defined as above. Then $f_{s'} = \overline{y}_{s'} - \overline{h}$ and it follows from (39) that $f_i \geq \overline{y}_i - \overline{h}$. Then $g_{s'i} \geq g_{s'i}(\overline{y}) = \overline{g}$. Similarly, when $t_\alpha < t'$, then $\overline{g} \leq g^-$.

The next easiest case is when $s' = s_\alpha$ and $t' = t_\alpha$. In this case,

$$g^+ = g^- = \frac{f_{t'} - f_{s'} - 2h}{x_{t'} - x_{s'}},$$

and it follows from (41) that this quantity is not less than $\bar{g}$, as required to establish (42).

It remains to resolve the two similar cases where $s' < s_\alpha$ and $t_\alpha = t'$, and where $s' = s_\alpha$ and $t_\alpha < t'$. It suffices to consider the former case and establish that $\bar{g} \leq g^-$. The procedure *closejoin* will add one or more indices $s'+, \ldots, i, s$ to I. It has already been shown that because $s' < s_\alpha$, $\bar{g} \leq g^+$. The gradients defined by the successive elements of I from $s'+$ to s will be monotonically increasing from $g_{s'} = g^+$, and so $g_{is} \geq g^+ \geq \bar{g}$. The gradient g^- is in this case the new join gradient. The remark in the paragraph after (25) then shows that $c_{s-1}(\mathbf{y}) \geq 0$, i.e. $g^- \geq g_{is}$ as required to establish (42). □

The proof of the effectiveness of Algorithm 4 and also of its generalization in the next section will require an important corollary of Lemma 5. When new sections are added on either side of an existing section, the convexity at the point where they are joined increases away from zero. Thus when Steps 2 to 6 of Algorithm 4 are applied, the constraints c_{k-1}, $c_{j^\star-1}$, and c_{k^+-1} increase away from zero, so that no more than two sign changes can be created in the second divided differences.

Corollary 4 *When defined, the constraints c_{k-1}, $c_{j^\star-1}$, and c_{k^+-1} satisfy the inequalities*

$$c_{k-1}(\mathbf{y}^2) \geq c_{k-1}(\mathbf{y}^0) \geq 0,$$

$$c_{j^\star-1}(\mathbf{y}^2) \leq c_{j^\star-1}(\mathbf{y}^0) \leq 0,$$

$$c_{k^+-1}(\mathbf{y}^2) \geq c_{k^+-1}(\mathbf{y}^0) \geq 0.$$

The effectiveness of Algorithm 4 can now be established.

Theorem 4 *Algorithm 4 produces a real number $h \geq 0$, a vertex set I, a piece set S of I, and a vector $\mathbf{y}^2 = \mathbf{y}(S)$ well defined by (37), (32) and (33) such that $\mathbf{y}^2 \in Y_2$ and $h = h(S) = F(\mathbf{y}^2) = \inf F(Y_2)$.*

Proof If Algorithm 4 stops in Step 1, then $h = h^+(1, n) = 0$, $I = I^+(1, n)$, and $S = \{n\}$. Then $h(S) = 0$. It follows from Theorem 2 that $\mathbf{f} \in Y_0^+ \subset Y_2$. Equation (37) will set $P_1 = [1, n]$ and (32) and (33) will set $\mathbf{y}^2 = \mathbf{f}$. The theorem is then immediately established.

Otherwise the quantities $j^\star$, k, and k^+ are well defined and satisfy the inequalities $1 \leq k < j^\star < k^+ \leq n$. (Each equality is possible, for example when $\mathbf{f} \in Y_1^\pm$.) It follows from Lemma 1 that at this point at the end of Step 1,

$$I = I^+(1, k) \cup I^+(k^+, n). \tag{46}$$

Step 2 is then entered. It inserts $j^\star$ into I, increases S to $\hat{S} = \{k, j^\star, n\}$, and calculates $\hat{h} = \max(h^+(1, k), h^+(k^+, n))$. Note that $j^\star$, k, and k^+ are now consecutive elements of I.

Steps 3 and 4 are then executed, calling *closejoin* in the ranges $[k, j^\star]$ and $[j^\star, k^+]$, in general increasing s_1, s_2, and h, and adding new elements to I. The new elements of S always satisfy the inequalities $k \leq s_1 \leq j^\star \leq s_2 \leq k^+$, and h cannot decrease, so $\hat{h} \leq h$. Note that s_2 can be increased to n, for example when $\mathbf{f} \in Y_1^+$.

Now consider the situation when the algorithm stops. If $h^{(1)} = 0$, the algorithm either stops in Step 4 when $h = 0$, or alternatively jumps straight from Step 5 to Step 6 re-calling *closejoin*(1) with $h > 0$. It follows from the properties of Algorithm 3 that when *closejoin* is called with $h > 0$, it cannot increase s to t. Therefore when the algorithm stops, if $h > 0$ then $s_1 < j^\star$ and $s_2 < k^+$. Now define t_1 and t_2 as for (37). Then $\mathbf{y}^2$ is well defined in all cases.

It must now be shown that $\mathbf{y}^2$ is feasible and optimal. The first main step is to establish (31), i.e.

$$I = I^+(1, s_1) \cup I^-(t_1, s_2) \cup I^+(t_2, n). \tag{47}$$

First consider the range $[1, s_1]$. Since $k \leq s_1$, $[1, s_1] = [1, k] \cup [k, s_1]$. Let $I_1 = I \cap [1, s_1]$. It follows from (46) and Theorem 3 that $I_1 = I^+(1, k) \cup I^+(k, s_1)$. It is trivial that $I_1 = I^+(1, s_1)$ unless $1 < k < s_1$. In this case there will exist neighbouring indices k^- and i of k in I such that $1 \leq k^- < k < i \leq s_1 \leq j^\star$. Let $h_0 = h^+(1, n)$ and $g_0 = g^{(1)}(\hat{S}, h_0)$. Since k^-, k, and k^+ are neighbours in $I^+(1, n)$, then $g_{k^-k} \leq g_{kk^+} = g_0$. Note that $g^{(2)}(\hat{S}, h_0) = g_0$. It is now possible to apply Lemma 5 successively with $\overline{h} = h_0$. The definition of $j^\star$ allows a first application of the lemma with $h' = \hat{h}$ and $S' = \hat{S}$ in the range $[k, j^\star]$ (i.e. with $\alpha = 1$) to infer that at entry to Step 4, $h^{(1)} \leq h_0$. This inequality and the definition of $j^\star$ then allow a second application of the lemma in the range $[j^\star, k^+]$, where also $\overline{g} = g_0$, to infer that $h \leq h_0$. If Step 6 is not entered, it follows from the first application of the lemma that $g_{ki} \geq g_0$. If *closejoin* is called again in Step 6, a third application of the lemma may be made, in the range $[k, j^\star]$, to yield that in this case also, $g_{ki} \geq g_0$. Then $g_{k^-k} \leq g_{ki}$. Lemma 2 can now be applied to prove that $I^+(1, k) \cup I^+(k, s_1) = I^+(1, s_1)$. Thus in all cases

$$I_1 = I^+(1, s_1).$$

In the same way, if $I_3 = I \cap [t_2, n]$, then $I_3 = I^+(t_2, n)$, trivially when $t_2 = n$ and otherwise by a single application of Lemma 5.

Now consider the range $[t_1, s_2]$. Let $I_2 = I \cap [t_1, s_2]$. Since $j^\star \in I$, t_1 can never exceed $j^\star$ and so $[t_1, s_2] = [t_1, j^\star] \cup [j^\star, s_2]$. In all cases $I_2 = I^-(t_1, j^\star) \cup I^-(j^\star, s_2)$, and it is trivial that $I_2 = I^-(t_1, s_2)$ unless $t_1 < j^\star < s_2$. In this case there will exist left and right neighbours of $j^\star$ in I. Denote them by j^- and j^+. Then the successive applications of Lemma 5 made above establish that $g_{j^-j^\star} \geq g_0$ and that $g_{j^\star j^+} \leq g_0$ so that Lemma 2 again applies to give that $I_2 = I^-(t_1, s_2)$. Equation (47) is then established.

The feasibility of $\mathbf{y}^2$ can now be proved. It is only necessary to examine the four (or possibly fewer) join constraints c_{s_1-1}, c_{t_1-1}, c_{s_2-1}, and c_{t_2-1} when $s_1 < t_1$ and $s_2 < t_2$. First consider the last two constraints. Since $s_2 < k^+$, Lemma 4 shows that $c_{s_2-1}(\mathbf{y}^2) \leq 0$ whenever s_2 exceeds $j^\star$. When $s_2 = j^\star$, the corollary to Lemma 5 applies to show that $c_{s_2-1}(\mathbf{y}^2) \leq 0$. Similarly, Lemma 4 when $t_2 < k^+$ and the corollary to Lemma 5 when $t_2 = k^+$ show that $c_{t_2-1}(\mathbf{y}^2) \geq 0$ whenever $t_2 < n$. When $t_2 = n$ the feasibility of $\mathbf{y}^2$ will follow automatically from the signs of the other constraints. Now consider the constraints c_{s_1-1}, c_{t_1-1}. When Step 6 is entered or $h^{(1)} = h$, the same reasoning applies to show that $c_{s_1-1}(\mathbf{y}^2) \geq 0$ when $s_1 > 1$ and $c_{t_1-1}(\mathbf{y}^2) \geq 0$ when $t_1 > 1$. (If $\mathbf{f} \in Y_0^-$, the algorithm will reduce t_1 to 1.) When Step 6 is not entered and $h > h^{(1)}$, Lemma 4 cannot be applied, but Lemma 5 and its corollary then show that feasibility is assured unless $s_1 > k$ or $s_1^+ < j^\star$. In these cases all the gradients calculated are well defined and the test in Step 5 allows the algorithm to stop only when $\mathbf{y}^2$ is feasible.

It is now necessary to show that $F(\mathbf{y}^2) = h$. The construction of $\hat{h}$ and h, the inequality $\hat{h} \leq h$, and (47) show that $h = h(S)$ and $|y_j - f_j| \leq h$ when $j \in [1, s_1] \cup [t_1, s_2] \cup [t_2, n]$, with equality when $j \in I$. For any other value of j in the range $[1, n]$, Lemma 3 and the inequality $h^{(1)} \leq h$ show that $|y_j - f_j| < h$ whether or not Step 6 is entered. Then $F(\mathbf{y}^2) = h$.

If $h = 0$, optimality is trivial. Otherwise $s_\alpha < t_\alpha$, for $\alpha = 1, 2$. Then it is possible to redefine $j^\star = j^\star(S)$ and to let k, k^+, and β be uniquely redefined from $j^\star$. Then $j^\star \in P_\beta$. Let $K = \{k, j^\star, k^+, s_1, t_1, s_2, t_2\}$. Then K cannot have fewer than five elements, even if $t_1 = s_2$. Assume first that $\beta = 1$. It follows by the same argument used in the proof of Theorem 3 that $c_{kj^\star k^+}(\mathbf{y}^2) = 0$, $c_{j^\star s_1 t_1}(\mathbf{y}^2) \geq 0$, and $c_{s_1 t_1 t_2}(\mathbf{y}^2) \leq 0$. Then if $F(\mathbf{v}) \leq h$, $c_{kj^\star k^+}(\mathbf{v}) < 0$, $c_{j^\star s_1 t_1}(\mathbf{v}) > 0$, and $c_{s_1 t_1 t_2}(\mathbf{v}) < 0$, and therefore the consecutive divided differences of $\mathbf{v}$ with indices in the subset K must change sign twice starting with a negative sign, so that by Theorem 1, $\mathbf{v} \notin Y_2$. If $\beta = 2$ or $\beta = 3$, the argument is similar. □

Clearly Algorithm 4 could easily be modified to calculate a best concave–convex–concave approximation, but it is easier to cover this case in the next subsection.

2.4 The General Case

The algorithm to be described constructs a best approximation to $\mathbf{f}$ over Y_q from $q + 1$ alternately convex and concave pieces joined where necessary by up to q straight line joins. These pieces are built up recursively from the pieces of a similar best approximation over Y_{q-2} by essentially the same method used in the previous section when $q = 2$. All the sections of the Y_{q-2} approximation remain in place except one determining the value of inf $F(Y_{q-2})$ which is deleted and replaced with a new piece of opposite convexity to the piece previously containing this section, and two new joins. The minimum value h of a best approximation over Y_q determined

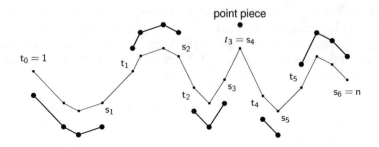

Fig. 6 Construction of Y_5 approximation

by the remaining sections is first calculated, then the procedure *closejoin* is called in each join. After this has been done, if h has increased then the resulting join constraints are checked and if necessary the calculation in each join is repeated with the new value of h. The construction of $\mathbf{y}$ is illustrated in Figure 6.

This method has the disadvantage that calculations sometimes have to be performed twice in each join, but each of the two sets of q join calculations can be performed in parallel. The procedure *closejoin* is a generalization of Algorithm 2, which is a modification of the method presented in Cullinan and Powell [3] which gave no proofs. That method avoids having to repeat itself by having an upper bound on h available throughout, but because of this does not admit of as much of its calculations being performed in parallel. It is therefore believed that the algorithm to be presented will often be more efficient.

Most of the notation needed for this case has already been developed. In particular, given a piece set S, (34) defines the corresponding price $h(S)$ and when this price is non-zero, (35) and (36) define the index $j^\star$ giving rise to it and the index β of the piece within which it lies. For any h the piece P_α is defined as $[t_{\alpha-1}, s_\alpha]$, where $t_0 = 1$, and for $\alpha \geq 1$ $t_\alpha = s_\alpha^+(I)$ when $h > 0$ and $t_\alpha = s_\alpha$ when $h = 0$. Since every element of I lies in P_α, for some α, the vector $\mathbf{y}(S)$ can be defined by

$$y_i = f_i + (-)^{\alpha-1}h \qquad \text{when } i \in I \cap P_\alpha, \qquad 1 \leq \alpha \leq q' + 1, \qquad (48)$$

$$y_j = y_j(I) \qquad 1 \leq j \leq n, \qquad (49)$$

where $q' = |S| - 1$ as before.

The following algorithm then calculates S, I, and h from which $\mathbf{y}^q$ is defined by these equations as the final value of $\mathbf{y}(S)$. When $h > 0$, the join gradients $g^{(\alpha)}(S, h)$ are defined by (38).

Algorithm 5 *To find a best Y_q approximation.*

Step 1.   Set $\bar{q} = q$ modulo 2.
            If $\bar{q} = 0$: set $S := \{n\}$ and use Algorithm 1 to calculate $I :=$
            $I^+(1, n)$ and $h = h^+(1, n)$.

Otherwise set $S := \{1, n\}$, $I := \{1, n\}$, $h := 0$ and call *closejoin*(1).

Step 2. If $\overline{q} = q$ or $h = 0$: **stop.**

Otherwise: set $j := j^\star(S)$, $k := j^-(I)$, insert j into I, and insert j and k into S, increase $\overline{q}$ by 2 and calculate $h := h(S)$. Set $h' := h$, $I' := I$, $S' := S$ and $\gamma := 1$.

Step 3. For $\alpha = 1$ to q: apply *closejoin*(α), and if h has increased set $\overline{\alpha} := \alpha$, and if $h = 0$ set $\gamma := \alpha + 1$. If $h = h'$: return to Step 2. Otherwise: set $\alpha = 1$ and go on to Step 4.

Step 4. If $\alpha < \gamma$ go to Step 5. If $\alpha = \overline{\alpha}$ return to Step 2. Calculate $g = g^{(\alpha)}(S, h)$. If $s_\alpha > s'_\alpha$ and $(-1)^{\alpha-1} g_{s_\alpha^-(I) s_\alpha} > (-1)^{\alpha-1} g$: go to Step 5. Set $t := s_\alpha^+(I')$ and $t^+ := t^+(I)$. If $t < s'^+_\alpha(I')$ and $(-1)^{\alpha-1} g_{t\, t^+} > (-1)^{\alpha-1} g$: go to Step 5. Increase α by 1 and repeat this step.

Step 5. Set $s_\alpha := s'_\alpha$, delete all elements of I lying strictly between s_α and $s_\alpha^+(I')$, and apply *closejoin*(α). Increase α by 1 and return to Step 4.

Step 1 initializes the two alternative first calculations when q is even and odd. Step 2 tests whether the calculation has come to an end, and if it has not the new values of $j^\star$ and β are calculated and recorded, as are the locations of the new pieces and the initial price. One final piece of bookkeeping is also prepared. To identify joins where a second call of *closejoin* will always be needed when h increases from zero, the index γ will be set by Step 3 to the lowest index for which $h > 0$ after the first call of *closejoin*(γ). Step 3 then performs the first q calls of *closejoin*, sets γ, and sets $\overline{\alpha}$ to the index of the call of *closejoin* in which h achieved its final value. If h has increased during this step, then Step 4 is entered. Tests are made to determine whether Step 5 will have to be entered to repeat the calculation in the join with index α, calculating a join gradient g only when necessary and possible. In particular it is never necessary to test once $\alpha \geq \overline{\alpha}$. In practice, in order to anticipate rounding errors, the tests whether $h = 0$ should be whether $h \leq 0$.

The applications of *closejoin* in each step could be performed simultaneously. In this case the largest of the ensuing values of the parameter h in Step 3 should be the value of h at entry to Step 4 and $\overline{\alpha}$ should be set to n. It will be shown in the proof of the following theorem that h is constant during Step 4.

The complexity of Algorithm 5 will be $qn \log n$.

Theorem 5 *Algorithm 5 produces a real number $h \geq 0$, an index set I, a piece set S of I, and a vector $\mathbf{y}^q = \mathbf{y}(S)$ well defined by (48) and (49) such that $\mathbf{y}^q \in Y_q$ and $h = F(\mathbf{y}^q) = \inf F(Y_q)$.*

Proof The proof that $\mathbf{y} \in Y_q$ and $F(\mathbf{y}) = h$ is by induction on $\bar{q}$. Assume that at any entry to Step 2 leading to Step 3, i.e. with $\bar{q} < q$ and $h > 0$, that

$$I = \bigcup_{\alpha=1}^{\bar{q}+1} I^{(-)^{\alpha-1}}(t_{\alpha-1}, s_\alpha), \tag{50}$$

where $t_0 = 1$ and $t_\alpha = s_\alpha^+(I)$ for $\alpha \geq 2$, and that

$$h = h(S). \tag{51}$$

The vector $\mathbf{y}(S)$ is then well defined. Assume that

$$F(\mathbf{y}(S)) = h \tag{52}$$

and that

$$\mathbf{y}(S) \in Y_{\bar{q}}. \tag{53}$$

It will be deduced from these equations that when the algorithm terminates $\mathbf{y}(S) \in Y_q$ and $h = \inf Y_q$.

It has been shown in Sections 2.1 and 2.2 that (50)–(52) hold at first entry to Step 2.

Most of the work needed for the proof has already been done in Lemmas 3–5. The main task is to examine the way h changes, so as to be able to apply these lemmas. Suppose that Step 2 begins with $h = \bar{h}$, $I = \bar{I}$ and $S = \bar{S}$, and that it increases $\bar{q}$ to q'. Step 2 also modifies S from $\bar{S}$ to $S' = \bar{S} \cup \{j, k\}$ and I from $\bar{I}$ to $I' = \bar{I} \cup \{j\}$, and recalculates h as $h' = h(S')$ using (34). Now by (51) and the definition of j in Step 2, $\bar{h} = h^\sigma(t_{\beta-1}, s_\beta)$, where β is defined in Step 2 by (36), $\sigma = (-)^{\beta-1}$, and $t_{\beta-1} = 1$ when $\beta = 1$ and $t_{\beta-1} = s_{\beta-1}^+(I)$ when $\beta \geq 2$. By Lemma 1, $I^\sigma(t_{\beta-1}, s_\beta) = I^\sigma(t_{\beta-1}, k) \cup I^\sigma(k^+, s_\beta)$, where $k^+ = k^+(\bar{I})$, and by definition, $h^\sigma(t_{\beta-1}, s_\beta) \geq \max(h^\sigma(t_{\beta-1}, k), h^\sigma(k^+, s_\beta))$. It follows that

$$h' \leq \bar{h}. \tag{54}$$

Furthermore, it follows from the definitions of $\bar{h}$ and β when $\alpha = \beta, \beta + 1$ and otherwise from (50)–(52) that

$$(x_j, f_j) \in \Pi_\alpha^{S'}(\bar{h}) \quad \text{for} \quad s_\alpha' \leq j \leq s_\alpha'^+(I').$$

Step 3 then applies *closejoin*(α) in each join. Let $h^{(\alpha)}$ be the value of h after *closejoin*(α) is called. Clearly the $h^{(\alpha)}$ are monotonically non-decreasing. Lemma 5 can now be applied successively beginning with (54) to show that

$$h^{(\alpha)} \leq \bar{h} \quad \text{for } 1 \leq \alpha \leq q'.$$

Let $t_0' = 1$ and $t_{\alpha-1}' = s_{\alpha-1}^+(I')$ when $\alpha \geq 2$. Then Lemma 5 and its corollary also show that whenever $t_{\alpha-1}' < s_\alpha' < s_\alpha$, the gradients on either side of $f_{s_\alpha'}$ have the correct monotonicity for Lemma 2 to yield that $I^\sigma(t_{\alpha-1}', s_\alpha') \cup I^\sigma(s_\alpha', s_\alpha) = I^\sigma(t_{\alpha-1}', s_\alpha)$, and that when $t_{\alpha-1} < t_{\alpha-1}' < s_\alpha'$ that $I^\sigma(t_{\alpha-1}, t_{\alpha-1}') \cup I^\sigma(t_{\alpha-1}', s_\alpha') = I^\sigma(t_{\alpha-1}, s_\alpha')$, where $\sigma = (-)^{\alpha-1}$. It follows either from this or otherwise trivially that in all cases after $closejoin(\alpha)$ is called, if P_α is defined from S then

$$I \cap P_\alpha = I^\sigma(t_{\alpha-1}, s_\alpha).$$

Because h cannot increase further after Step 3, this holds true whether $closejoin(\alpha)$ is called once only or again in Step 5, so that when Step 2 is next entered,

$$I = \bigcup_{\alpha=1}^{q'+1} I^{(-)^{\alpha-1}}(t_{\alpha-1}, s_\alpha). \tag{55}$$

The next step is to establish that when Step 2 is next entered with $h > 0$ then $h = h(S)$. Since $\bar{h} > 0$ and $k < j < k^+$ in Step 2, the quantity $h(S)$ to which h is set by Step 2 is well defined by (34) with $t_\alpha = s_\alpha^+(I)$. Thus

$$h' = h(S').$$

If Step 5 recalls $closejoin(\alpha)$ for any α, it will not increase h further, so h always attains its final value by the end of Step 3. It must be shown that the quantity $h(S)$ is always well defined when Step 2 is next entered. This can only fail to be the case if there is an α for which $h^{(\alpha)} = 0$. In such a case it must hold that $h' = 0$ and that h increased during Step 3. Then $\bar{\alpha}$ will be set to the index of the call of $closejoin$ in which h achieved its final positive value, and the parameter γ will be set to the lowest index for which $h^{(\gamma)} > 0$. Clearly, then, $\gamma \leq \bar{\alpha}$ and $\alpha < \gamma \leq \bar{\alpha}$, so that Step 4 will branch to Step 5, recalling $closejoin(\alpha)$ with a positive h so that afterwards $t_\alpha = s_\alpha^+(I)$. The quantity $h(S)$ will then be well defined when Step 2 is entered. It follows from the form of Algorithm 2 and (55) that whether h increases from h' or not, then when Step 2 is next entered,

$$h = h(S). \tag{56}$$

The vector $\mathbf{y} = \mathbf{y}(S)$ will then be well defined at the next entry to Step 2. The next argument will establish that (52) then holds. It follows from (55) and (56) that it is sufficient to show that for any indices j such that $s_\alpha < j < t_\alpha$, $|y_j(S) - f_j| < h$. This is equivalent to the statement that the graph points $(x_j, y_j(S))$ lie inside the parallelogram $\Pi_\alpha^S(h)$. It follows from Lemma 3 that immediately after $closejoin(\alpha)$ is last applied, $(x_j, y_j(S))$ lies strictly within $\Pi_\alpha^S(h)$. If this call of $closejoin$ comes in Step 5, h has already attained its value at next entry to Step 2 and the result is immediately established. If, on the other hand, this call of $closejoin$ comes in Step 3, then $(x_j, y_j(S))$ lies inside $\Pi_\alpha^S(h^{(\alpha)})$, and $h^{(\alpha)} \leq h$. A simple calculation shows that if $a < b$ then $\Pi_\alpha^S(a) \subset \Pi_\alpha^S(b)$. The result then follows. Thus at next entry to Step 2,

$$F(\mathbf{y}(S)) = h.$$

The next argument will re-establish (53) at that point. It follows from (50) that it is only necessary to establish that the join constraints $c_{s_\alpha-1}$ and $c_{t_\alpha-1}$ of $\mathbf{y}$ have the correct signs when $2 \le s_\alpha < t_\alpha \le n-1$, i.e. when $h > 0$.

When $1 < s_\alpha = s'_\alpha$, (53) and the corollary to Lemma 5 imply that $c_{s'_\alpha-1}(\mathbf{y})$ had the correct sign at the last exit from Step 2 and has not now moved closer to zero. The same is true of $c_{t_\alpha-1}$ when $t_\alpha = t'_\alpha < n$. When $s_\alpha > s'_\alpha$, Lemma 4 applies provided that h has not increased after the last call of $closejoin(\alpha)$ to yield that $(-1)^{\alpha-1}c_{s_\alpha-1} \ge 0$ and that when $1 < t_\alpha < t'_\alpha$ that $(-1)^{\alpha-1}c_{t_\alpha-1} \le 0$. If $closejoin(\alpha)$ is only called once and h increases subsequently, it must be shown that the test in Step 4 is adequate to ensure feasibility. In this case Step 4 will certainly be entered (because $h > h'$) and Step 3 will set $\bar{\alpha}$ and γ so that $\gamma \le \alpha < \bar{\alpha}$, so that $h^{(\alpha)} > 0$. It follows that $s_\alpha < t_\alpha$ and so g is well defined and the correct sign of $c_{s_\alpha-1}$ is assured by the test in Step 5. The case $t_\alpha < t'_\alpha$ is similar. Thus at next entry to Step 2, even when $h = 0$,

$$\mathbf{y} \in Y_{q'}.$$

Thus (50)–(53) are established by induction. It follows that when the algorithm terminates with a number h and the pieces from which the vector $\mathbf{y}$ is constructed, $\mathbf{y} \in Y_{\bar{q}} \subset Y_q$ and $F(\mathbf{y}) = h$.

The proof of optimality is similar to that in Theorem 4. If $h = 0$, there is nothing to prove. Otherwise, once again a set K is constructed containing $q+3$ indices of data points on alternating sides of the components of $\mathbf{y}$ and all the same distance h from them, and such that the consecutive divided differences with indices in K of any vector $\mathbf{v} \in \mathbb{R}^n$ giving a lower value of F than h change sign q times starting with a negative sign, so that by Theorem 1, $\mathbf{v} \notin Y_q$. As in the case $q = 2$ there seems no straightforward way of constructing an appropriate set with exactly $q+3$ elements. Therefore, after Algorithm 5 has terminated, let $s_\alpha, t_\alpha, 1 \le \alpha \le q+2$, be defined as the join points that would next have been constructed in Step 2 if the algorithm had not terminated (i.e. by adding k, $j^\star$, and k^+ to the existing set of join points and reindexing, but without increasing q.) Then $s_\beta = k, t_\beta = s_{\beta+1} = j^\star$, and $t_{\beta+1} = k^+$. It follows as in the proof of Theorem 3 that because $h > 0$, $s_\alpha < t_\alpha$ for all α. Now define $K = \{s_\alpha, t_\alpha : 1 \le \alpha \le q+2\}$. Because the piece $[t_\beta, s_{\beta+1}]$ has only one element and any of the other pieces can have only one element, this set K can have from $q+3$ to $2q+3$ elements.

It follows from the construction of $\mathbf{y}$ and (53) that $(-1)^{\alpha-1}c_i(\mathbf{y}) \le 0$ for $s_\alpha \le i \le t_{\alpha+1}-2$ and so, from the corollary to Theorem 1, that for any α in the range $1 \le \alpha \le q+2$, $(-1)^{\alpha-1}c_{s(\alpha),t(\alpha),t(\alpha+1)}(\mathbf{y}) \le 0$ and $(-1)^{\alpha-1}c_{s(\alpha),s(\alpha+1),t(\alpha+1)}(\mathbf{y}) \le 0$. From (48),

$$y_{s_\alpha} = f_{s_\alpha} + (-1)^{\alpha-1}h$$

$$y_{t_\alpha} = f_{t_\alpha} - (-1)^{\alpha-1}h.$$

If $t_\alpha = s_{\alpha+1}$, it follows immediately from these equations that if $F(\mathbf{v}) < h$

$$(-1)^{\alpha-1} c_{s(\alpha),t(\alpha),t(\alpha+1)}(\mathbf{v}) < 0.$$

Otherwise when $t_\alpha < s_{\alpha+1}$, it follows from these equations that if $F(\mathbf{v}) < h$, then $(-1)^{\alpha-1} c_{s(\alpha),t(\alpha),t(\alpha+1)}(\mathbf{v}) < 0$ and $(-1)^{\alpha-1} c_{s(\alpha),s(\alpha+1),t(\alpha+1)}(\mathbf{v}) < 0$ and hence, from the corollary to Theorem 1, that the *consecutive* divided differences with indices in K satisfy

$$(-1)^{\alpha-1} c_{s(\alpha),t(\alpha),s(\alpha+1)}(\mathbf{v}) < 0 \quad \text{or} \quad (-1)^{\alpha-1} c_{t(\alpha),s(\alpha+1),t(\alpha+1)}(\mathbf{v}) < 0.$$

Therefore in all cases the divided differences of $\mathbf{v}$ with indices in K have at least q sign changes starting with a negative sign, and so if $\mathbf{v}$ is any vector for which $F(\mathbf{v}) < h$, then $\mathbf{v} \notin Y_q$. □

Thus a global solution to the optimization problem of this section has been constructed in all cases.

3 Numerical Results and Conclusions

This section will describe the results of some tests of the data smoothing method developed in the previous section. Algorithm 5 was trivially extended to find a best approximation over $Y_{\pm q}$. It was coded in PASCAL and run using Lazarus Pascal Version 1.0.8 on a desktop PC.

3.1 Synthetic Test Data

The tests were almost all conducted by contaminating sets of values of a known function with errors, applying the method, and then comparing the results with the exact function values. The errors added to exact function values were either truncation or rounding errors, or uniformly distributed random errors in the interval $[-\epsilon, \epsilon]$.

If $\mathbf{g}$ is the vector of exact function values, one simple measure of the effectiveness of the method can be obtained from the quantity

$$P_p = \left(1 - \frac{\|\mathbf{y} - \mathbf{g}\|_p}{\|\| - \mathbf{g}\|_p}\right),$$

obtained from the ℓ_p norm. For each set of data the values of P_∞ and P_2 were calculated.

Table 1 The zero function

n	P_∞	P_2
5001	-19.37	94.51
501	-68.97	76.25

As a preliminary test, equally spaced values of the zero function on $[-5, 5]$ were contaminated with uniformly distributed errors with $\epsilon = 0.1$. The results are shown in Table 1.

The difference between y_j and zero was of the order of 10^{-4} for most of the range, but near the ends it rose to 10^{-1}, accounting for the relatively high value $\|\mathbf{y} - \mathbf{g}\|_2 = 0.3098$ when $n = 501$. High end errors were frequent and so the value of P_∞ was not usually a reliable indicator of the efficiency of the method. These errors can occur because if, for example, $\sigma = +$, then $y_1 = f_1 + h$ and if $f_1 = g_1 + \epsilon$ then $\|\mathbf{y} - \mathbf{g}\|_\infty \geq h + \epsilon = \|\mathbf{y} - \mathbf{f}\|_\infty + \epsilon$.

Two types of data were considered next: undulating data and peaky data, because these types of data can be hard to smooth using divided differences unless sign changes are allowed. The first category of data were obtained from equally spaced values of the function $\sin \pi x$ on the interval $[-2, 2]$, and the second from equally spaced values of the normal distribution function $N_s(x) = (2\pi x)^{-\frac{1}{2}} \exp(-x^2/(2s^2))$ with $s = 0.8$, on the interval $[-5, 5]$. These same functions were previously used to test the ℓ_2 data smoothing method in Cullinan [2] which did not allow sign changes, and it was shown there that the sine data were possible to treat well but the peaked data did not give very good results.

Many of the results obtained using Algorithm 5 looked very acceptable when graphed, even when P_2 was only moderate. The results of Table 1 show why it can be quite difficult for P_2 to approach 100 even when the results are very good, partly because of end errors, but also because of another phenomenon that can move smoothed points further away from the underlying function values. The method raises convex pieces and lowers concave pieces, and this often results in points near an extremum of the underlying function being pulled away from it, for example if there is a large error on the low side of a maximum.

Different sets of random errors with the same ϵ can give very different values of P_2, particularly when the spacing between points is not very small, and in fact it was found that reducing the spacing beyond a certain amount can make a great difference to the consistency of the results.

The method coped quite well with sine data. For example, with $f_j = \sin 2\pi x_j + \epsilon \sin 1000 x_j$ and equally spaced data on the interval $[-2, 0]$, when $\epsilon = 0.02$ and $n = 101$, $P_2 = 10.2$, when $n = 1001$, $P_2 = 80.42$, when $n = 10001$, $P_2 = 93.01$, and when $n = 100001$, $P_2 = 93.34$. With $\epsilon = 0.1$ the respective figures for P_2 were 56, 87.82, 93.05, and 93.38. The method was also run with these sine data but at variable abscissæ defined by replacing the equally spaced ones x_j with $x_j + 0.25h \sin 100000 x_j$. The timings did not increase even with $n = 100001$ and the values of P_2 were comparable. The method was also very good with the peaked data, and an improvement on the method in Cullinan [2], managing to model both the peak and the flat tail well. The results of one run are shown in Figure 7.

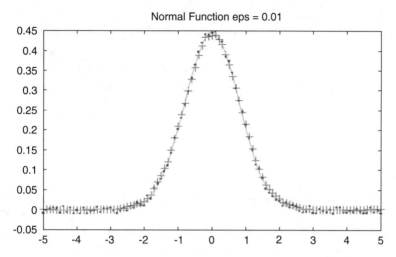

Fig. 7 Peaked data with $n = 101$, $q = 2$, $\epsilon = 0.01$, $P_2 = 45.8$

In all the above cases the best choice of q was obvious. The next trials examined two functions where it was not. The first was the first three terms of the Fourier series for a square wave and the last was a wiggly function requiring a high value of q.

A good strategy when the choice of q is not obvious and when there is some bound on the expected errors is to run the algorithm with increasing values of q until the value of h decreases to a value of the same order as the expected errors. This is because of the following result. Suppose that $\sigma = +$, the Algorithm finishes with $h > 0$, and that $\|\mathbf{g} - \mathbf{f}\|_\infty = \epsilon$. Let k, $j^\star$, k^+ be determined by the Algorithm and suppose that $c_{kj^\star k^+}(\mathbf{g}) \geq 0$. Then

$$h \leq \epsilon. \tag{57}$$

The proof is straightforward. Since $\|\mathbf{g} - \mathbf{f}\|_\infty = \epsilon$, then $f_k \geq g_k - \epsilon$, $f_{j^\star} \leq g_{j^\star} + \epsilon$, $f_{k^+} \geq g_{k^+} - \epsilon$, and $c_{kj^\star k^+}(\mathbf{g}) \geq 0$. Define quantities $z_k = -1$, $z_{j^\star} = 0$, $z_{k^+} = -1$ and note that $z_{j^\star} - z_{j^\star}(k, k^+) = 2$. The proof then relies on (9), the construction of h from f_k, $f_{j^\star}$, and f_{k^+}, and $c_{kj^\star k^+}$ being an increasing linear function of its first and last arguments and a decreasing linear function of its middle one. Then $c_{kj^\star k^+}(\mathbf{f}) = -2(x_{j^\star} - x_k)(x_{k^+} - x_{j^\star})h \geq c_{kj^\star k^+}(\mathbf{g}) + \epsilon c_{kj^\star k^+}(\mathbf{z}) = c_{kj^\star k^+}(\mathbf{g}) - 2(x_{j^\star} - x_k)(x_{k^+} - x_{j^\star})\epsilon$ and so $h \leq \epsilon$.

It follows from this result that once q is large enough for the points k, $j^\star$, k^+ that partly determine h to lie in the same piece of g then there is an upper bound on h. Thus a suitable value for q can be found by running the algorithm with a high value of q outputting h just before the algorithm increases $\bar{q}$ each time until h decreases to within the expected error. If this is not economic because of the size of n then the algorithm could perhaps be applied with high q to a uniformly distributed subset of the data until a suitable value is found.

Table 2 The wiggly function
for increasing n

n	h	P_2	Time (ms)
1001	0.001700	72.82	6
10,001	0.001976	85.35	14
100,001	0.001978	85.53	129
1,000,001	0.001979	85.54	1940
2,000,001	0.001979	85.54	2881
5,000,001	0.001979	85.54	10,279
10,000,001	0.001979	85.54	21,183

In the case of the square wave approximation function $g(x) = \frac{4}{\pi}(\sin\frac{1}{2}\pi x +$
$\frac{1}{3}\sin\frac{3}{2}\pi x + \frac{1}{5}\sin\frac{5}{2}\pi x)$ it is hard to see from the graph whether the first piece is
convex or concave when $q = 4$. The algorithm was run with $f_j = g_j + \epsilon \sin 1000x_j$,
$\epsilon = 0.02$ and 1001 equally spaced data on $[-1, 1]$. When $q = 4$, $h = 0.151$ but
when $q = 5$ then $h = 0.0197$. Increasing q further to 5 and 9 did not reduce h any
more.

The last constructed data to be tested came from the 'wiggly' function $g(x) =$
$\cos \pi x - 0.3 \cos 5\pi x - 0.2 \sin 20\pi x$ and $f_j = g_j + \epsilon \sin(1000(x_j + 1))$, which
has many local maxima (illustrated below). The original choice of perturbation used
the function $\sin 1000x$ as above but when the algorithm was run with equally spaced
data on the interval $[-2, 0]$ on the unperturbed data with $n = 101$ and $q = 2$ it failed
for the reason given after the description of Algorithm 2. This happened because the
four unperturbed data points with abscissæ $-0.78, -0.62, -0.38, -0.224$ formed a
parallelogram. Both $g(x)$ and the perturbing function were antisymmetric and so the
perturbed data would also cause the algorithm to fail. Therefore a non-symmetric
perturbation function was used instead and no further failures were encountered.
The kind of symmetry that leads to this failure seems unlikely in practice but should
it occur the remedy proposed is to perturb one of the data causing the problem
by a tiny amount. In the above case perturbing the datum with abscissa -0.62 by
0.5×10^{-11} cured the problem.

The data defined in the last paragraph were next used to test the Algorithm for
large n. With small n it was found that the value of h reduced acceptably for $q = 40$.
The algorithm was then run with equally spaced data on the interval $[-1, 1]$ with
$\epsilon = 0.02$, $q = 40$ and n increasing from 1001 to 10,000,001. Some of the results
are shown in Table 2.

These timings are not claimed to be optimal but they do seem consistent with the
Algorithm being $o(n \log n)$.

The ability of the Algorithm to cope with large errors is illustrated in Figure 8.

3.2 Real Test Data

In addition to the test data got from a known underlying function, one case of real
empirical data was examined. This was a set of 579 measurements of deuterium

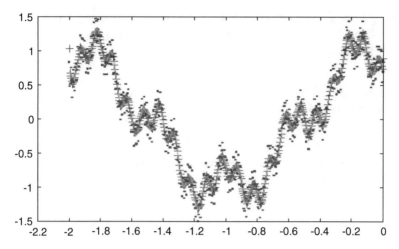

Fig. 8 Wiggly function with $n = 501$, $q = 40$, $\epsilon = 0.2$, $P_2 = 79.6$

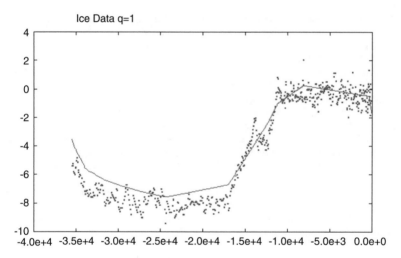

Fig. 9 Vostok ice data with $q = 1$

content in Vostok ice cores over a very long period. The set was kindly provided by I. Demetriou from a much larger set used to test his data-smoothing algorithms and originally provided by Petit et al. [10]. These data are not ideally suited to minimax approximation because they appear to have large errors and quite a few outliers. It was not clear from the graph of the data whether to run the method with $q = 1$ or with $q = 3$. The results of a run with $q = 1$ are illustrated in Figure 9. The method did not model the left-hand part of the range very well and exhibits an end error at the right-hand end. One strategy that can be adopted in cases with large outliers is to delete or re-weight them. This strategy has in fact been used by Demetriou.

Deleting a few obvious outliers was tried and applying the method to the resulting dataset did model the left-hand part better, but neither weighting nor deletion seems an ideal choice with respect to minimax approximation.

3.3 Conclusion

It can be said that this method is very economical, and can give very good results, particularly when the data are close together. Therefore it seems a particularly good candidate for smoothing large densely packed data with uniform errors, even when the errors are relatively large.

References

1. Cullinan, M.P.: Data smoothing by adjustment of divided differences. Ph.D. dissertation, Department of Applied Mathematics and Theoretical Physics, University of Cambridge (1986)
2. Cullinan, M.P.: Data smoothing using non-negative divided differences and ℓ_2 approximation. IMA J. Numer. Anal. **10**, 583–608 (1990)
3. Cullinan, M.P., Powell, M.J.D.: Data smoothing by divided differences. In: Watson, G.A. (ed.) Numerical Analysis Proceedings, Dundee 1981. Lecture Notes in Mathematics, vol. 912, pp. 26–37. Springer, Berlin (1982)
4. Demetriou, I.C.: Signs of divided differences yield least squares data fitting with constrained monotonicity or convexity. J. Comput. Appl. Math. **146**(2), 179–211 (2002)
5. Demetriou, I.C. Powell, M.J.D.: Least squares fitting to univariate data subject to restrictions on the signs of the second differences. In: Iserles, A. Buhmann, M.D. (eds.) Approximation Theory and Optimization. Tributes to M. J. D. Powell, pp. 109–132. Cambridge University Press, Cambridge (1997)
6. Graham, R.L.: An efficient algorithm for determining the convex hull of a finite point set. Inform. Proc. Lett. **1**, 132–133 (1972)
7. Hildebrand, F.B.: Introduction to Numerical Analysis. McGraw Hill, New York (1956)
8. Kruskal, J.B.: Non-metric multidimensional scaling: a numerical method. Psychometrika **29**, 115–129 (1964)
9. Miles, R.E.: The complete amalgamation into blocks, by weighted means, of a finite set of real numbers. Biometrika **46**, 317–327 (1959)
10. Petit, J.R., et al.: Vostok Ice Core Data for 420,000 Years, IGBP PAGES/World Data Center for Paleoclimatology Data Contribution Series #2001-076. NOAA/NGDC Paleoclimatology Program, Boulder (2001)
11. Ubhaya, V.A.: An $O(n)$ algorithm for discrete n-point convex approximation with applications to continuous case. J. Math. Anal. Appl. **72**(1), 338–354 (1979)

A Decomposition Theorem for the Least Squares Piecewise Monotonic Data Approximation Problem

Ioannis C. Demetriou

Abstract We consider the problem of calculating the least squares approximation to n measurements that contain random errors of function values subject to the condition that the first differences of the approximated values have at most $k - 1$ sign changes, where k is a given positive integer. The positions of the sign changes are integer variables whose optimal values are to be determined automatically. Since the number of trials of all possible combinations of positions in order to find an optimal one is of magnitude n^{k-1}, it would not be practicable to consider each one separately. We give a characterization theorem which shows that the problem reduces to partitioning the data into at most k disjoint sets of adjacent data and solving a $k = 1$ problem for each set (monotonic approximation case). The important computational consequence of this theorem is that it allows dynamic programming to be applied for obtaining the partition and solving the whole problem in only a quadratic number of operations. However, shorter computation times in practice are confirmed by our numerical results. Further, an example is given, which shows that the time required by the dynamic programming method to locate optimally peaks when $k = 50$ in a NMR spectrum that consists of about 110,000 data points is less than a minute, but the number of trials of all possible combinations would be of magnitude 10^{250}.

1 Introduction

A characterization theorem is presented for the following data approximation problem, which has been introduced by Demetriou and Powell [10]. The solution to the problem is known, but the theorem provides necessary and sufficient conditions in a unified form. Let $\{\phi_i : i = 1, 2, \ldots, n\}$ be measurements of the real function

I. C. Demetriou (✉)
Department of Economics, University of Athens, Athens, Greece
e-mail: demetri@econ.uoa.gr

© Springer Nature Switzerland AG 2019
I. C. Demetriou, P. M. Pardalos (eds.), *Approximation and Optimization*,
Springer Optimization and Its Applications 145,
https://doi.org/10.1007/978-3-030-12767-1_7

values $\{f(x_i) : i = 1, 2, \ldots, n\}$, where the abscissae $\{x_i : i = 1, 2, \ldots, n\}$ are in strictly ascending order. If the measurements are contaminated by random errors, then it is likely that the sequence of first differences $\{\phi_{i+1} - \phi_i : i = 1, 2, \ldots, n-1\}$ contains far more sign changes than the sequence $\{f(x_{i+1}) - f(x_i) : i = 1, 2, \ldots, n-1\}$. Therefore, for some integer k that is much smaller than n, we seek numbers $\{y_i : i = 1, 2, \ldots, n\}$ that make the least sum of squares change to the measurements so that the sequence $\{y_{i+1} - y_i : i = 1, 2, \ldots, n-1\}$ changes sign at most $k - 1$ times. We regard the original measurements and the approximated data as n-vectors $\boldsymbol{\phi}$ and $\mathbf{y}$. The constraints on $\mathbf{y}$ allow at most k sections of monotonic components, alternately increasing and decreasing. Without loss of generality we suppose that the first monotonic section is increasing.

Hence we denote by $Y(k, n)$ the set of n-vectors $\mathbf{y}$ whose components satisfy the piecewise monotonicity constraints

$$\left. \begin{array}{l} y_{t_{j-1}} \leq y_{t_{j-1}+1} \leq \cdots \leq y_{t_j}, \ j \text{ is odd} \\ y_{t_{j-1}} \geq y_{t_{j-1}+1} \geq \cdots \geq y_{t_j}, \ j \text{ is even} \end{array} \right\}, \tag{1}$$

where the integers $\{t_j : j = 1, 2, \ldots, k - 1\}$ satisfy the conditions

$$1 = t_0 \leq t_1 \leq \cdots \leq t_k = n, \tag{2}$$

and the optimization calculation seeks a vector $\mathbf{y}^*$ in $Y(k, n)$ that minimizes the sum of squares of residuals

$$\|\mathbf{y} - \boldsymbol{\phi}\|_2^2 = \sum_{i=1}^{n} (y_i - \phi_i)^2. \tag{3}$$

Such a vector $\mathbf{y}^*$ is called an optimal piecewise monotonic approximation to $\boldsymbol{\phi}$. Since $\{t_j : j = 1, 2, \ldots, k - 1\}$ are also variables of the optimization problem, there are $O(n^{k-1})$ combinations of these integers in order to find a combination that gives an optimal approximation, which can make an exhaustive search prohibitively expensive.

In Section 2 we present a theorem that proves a decomposition property of the problem. The important consequence of this decomposition is that an optimal approximation can be generated by a dynamic programming procedure. Some algorithms of this kind are given by Demetriou [6, 9] and Demetriou and Powell [10]. Depending on the implementation, they obtain a solution in at most $O(kn^2)$ or $O(n^2 + kn \log_2 n)$ computer operations. Next the typical amount of computation times in practice is demonstrated by some numerical results. In Section 3, the efficacy of the technique is illustrated by an application for locating peaks in an NMR spectrum that consists of 110,403 pairs of data. This problem arises in the practice of spectroscopy calculations and is of intrinsic interest. In Section 4 we give a brief summary.

Two related calculations are studied by Demetriou [5, 8] and Cullinan and Powell [3], which instead of (3) minimize the sum of moduli $||\mathbf{y} - \boldsymbol{\phi}||_1 = \sum_{i=1}^{n} |y_i - \phi_i|$ and the supremum norm $||\mathbf{y} - \boldsymbol{\phi}||_\infty = \max_{1 \leq i \leq n} |y_i - \phi_i|$, respectively, subject to the same constraints on $\mathbf{y}$.

2   The Theorem

By taking advantage of some properties that depend on the approximation problem and are not obtained generally when one seeks values of the integer variables $\{t_j : j = 1, 2, \ldots, k-1\}$, the optimization calculation that is stated in Section 1 reduces to partitioning the data into at most k disjoint sets of adjacent data and solving a $k = 1$ problem for each set. These properties are given by Demetriou and Powell [10] and in brief they are as follows.

I. If both $\mathbf{y}$ and the integer variables $\{t_j : j = 0, 1, \ldots, k\}$ are optimal, then, provided that $\boldsymbol{\phi}$ is not in $Y(k, n)$, these integer variables are all different.

II. The component $y_{t_j}, 1 \leq j \leq k-1$ of an optimal $\mathbf{y}$ is independent of $\{y_i : i \neq t_j\}$, which implies the interpolation equations

$$y_{t_j} = \phi_{t_j}, \ j = 1, 2, \ldots, k-1. \tag{4}$$

III. If $\mathbf{y}$ is optimal and if this approximation has a monotonic increasing section on $[t_{j-1}, t_j]$, then the components $y_i, i = t_{j-1}, t_{j-1} + 1, \ldots, t_j$ have the values that minimize the sum of squares $\sum_{i=t_{j-1}}^{t_j} (y_i - \phi_i)^2$ subject to the constraints $y_i - y_{i+1} \leq 0, i = t_{j-1}, t_{j-1} + 1, \ldots, t_j - 1$. Similarly, if $\mathbf{y}$ has a monotonic decreasing section on $[t_{j-1}, t_j]$, then the components $y_i, i = t_{j-1}, t_{j-1} + 1, \ldots, t_j$ have the values that minimize the sum of squares $\sum_{i=t_{j-1}}^{t_j} (y_i - \phi_i)^2$ subject to the constraints $y_i - y_{i+1} \geq 0, i = t_{j-1}, t_{j-1}+1, \ldots, t_j - 1$. Therefore if one knows the optimal integer variables $\{t_j : j = 1, 2, \ldots, k-1\}$, then the components of $\mathbf{y}$ are obtained by solving these particular problems separately between adjacent optimal integer variables. Hence it helps our analysis to introduce the notation

$$\alpha(t_{j-1}, t_j) = \min_{y_{t_{j-1}} \leq y_{t_{j-1}+1} \leq \cdots \leq y_{t_j}} \sum_{i=t_{j-1}}^{t_j} (y_i - \phi_i)^2, \ \text{if } j \text{ is odd}, \tag{5}$$

and

$$\beta(t_{j-1}, t_j) = \min_{y_{t_{j-1}} \geq y_{t_{j-1}+1} \geq \cdots \geq y_{t_j}} \sum_{i=t_{j-1}}^{t_j} (y_i - \phi_i)^2, \ \text{if } j \text{ is even}. \tag{6}$$

We note that the constraints that occur in (5) on the components $\{y_i : i = t_{j-1}, t_{j-1} + 1, \ldots, t_j\}$ are linear with linearly independent normals. Further, the second derivative matrix of the objective function $\sum_{i=t_{j-1}}^{t_j} (y_i - \phi_i)^2$ is twice the $(t_j - t_{j-1}) \times (t_j - t_{j-1})$ unit matrix. Therefore (5) is a strictly convex quadratic programming problem that has a unique solution, which we call best monotonic increasing approximation to $\{\phi_i : i = t_{j-1}, t_{j-1} + 1, \ldots, t_j\}$. The solution can be calculated by the special algorithms of Cullinan and Powell [3] and Demetriou and Powell [10] that are based on van Eeden's method [13]. These algorithms are far more efficient than general quadratic programming algorithms (for a general reference see Fletcher [11]). Moreover, Algorithm 1 of [10] computes $\{y_i : i = t_{j-1}, t_{j-1}+1, \ldots, t_j\}$ and all the numbers $\alpha(t_{j-1}, i)$, for $i = t_{j-1}, t_{j-1}+1, \ldots, t_j$) in only $O(t_j - t_{j-1})$ computer operations, which is highly suitable for the needs of our computation.

In order to present the theorem that is mentioned in Section 1, we need some extra notation. Let $k > 1$, let $T = \{t_0, t_1, \ldots, t_k\}$ be a set of integers that satisfy the conditions (2) and let $\mathbf{y}(T)$ be the n-vector $\mathbf{y}$ that minimizes the objective function (3) subject to the constraints that, for $j = 1, 2, \ldots, k$ the sequence $\{y(T)_i : i = t_{j-1}, t_{j-1} + 1, \ldots, t_j\}$ is monotonic increasing if j is odd and monotonic decreasing if j is even. Thus, $\mathbf{y}(T)$ is the unique solution of a strictly convex quadratic programming calculation.

Theorem 1 *The vector* $\mathbf{y}(T)$ *minimizes the objective function* (3) *subject to the constraints* $\mathbf{y} \in Y(k, n)$ *if and only if the equation*

$$\sum_{j=1,\ j\ odd}^{k} \alpha(t_{j-1}, t_j) + \sum_{j=1,\ j\ even}^{k} \beta(t_{j-1}, t_j) =$$
$$\min_{1=s_0 \leq s_1 \leq \cdots \leq s_k = n} \left\{ \sum_{j=1,\ j\ odd}^{k} \alpha(s_{j-1}, s_j) + \sum_{j=1,\ j\ even}^{k} \beta(s_{j-1}, s_j) \right\} \quad (7)$$

holds.

Proof We consider first the necessary conditions. We prove that if $\mathbf{y}(T)$ minimizes (3) subject to $\mathbf{y} \in Y(k, n)$, then Equation (7) is obtained. In view of property II, the sequence $\{y(T)_i : i = t_{j-1}, t_{j-1} + 1, \ldots, t_j\}$ is the best monotonic increasing approximation to the data $\{\phi_i : i = t_{j-1}, t_{j-1}+1, \ldots, t_j\}$ if j is odd and it is the best monotonic decreasing approximation if j is even, because otherwise we can reduce $\|\mathbf{y}(T) - \boldsymbol{\phi}\|_2$ by replacing $\{y(T)_i : i = t_{j-1}, t_{j-1} + 1, \ldots, t_j\}$ by the best monotonic approximation to the data $\{\phi_i : i = t_{j-1}, t_{j-1} + 1, \ldots, t_j\}$, which preserves $\mathbf{y}(T) \in Y(k, n)$. Hence we have

$$\sum_{i=t_{j-1}}^{t_j} (y_i - \phi_i)^2 = \begin{cases} \alpha(t_{j-1}, t_j), & j\ odd \\ \beta(t_{j-1}, t_j), & j\ even. \end{cases} \quad (8)$$

We add (8) for $j = 1, 2, \ldots, k$, we take into account (4) and we see that the left-hand side of the expression (7) has the value $||\mathbf{y}(T) - \boldsymbol{\phi}||_2^2$, namely

$$\sum_{j=1,\ j\ \text{odd}}^{k} \alpha(t_{j-1}, t_j) + \sum_{j=1,\ j\ \text{even}}^{k} \beta(t_{j-1}, t_j) = ||\mathbf{y}(T) - \boldsymbol{\phi}||_2^2. \tag{9}$$

Consequently the value $||\mathbf{y}(T) - \boldsymbol{\phi}||_2^2$ implies the bound on the right-hand side of (7),

$$\min_{1=s_0 \le s_1 \le \cdots \le s_k = n} \left\{ \sum_{j=1,\ j\ \text{odd}}^{k} \alpha(s_{j-1}, s_j) + \sum_{j=1,\ j\ \text{even}}^{k} \beta(s_{j-1}, s_j) \right\} \le ||\mathbf{y}(T) - \boldsymbol{\phi}||_2^2. \tag{10}$$

It follows that the Equation (7) is satisfied, provided that we can establish the inequality

$$||\mathbf{z}^* - \boldsymbol{\phi}||_2^2 \le \min_{1=s_0 \le s_1 \le \cdots \le s_k = n} \left\{ \sum_{j=1,\ j\ \text{odd}}^{k} \alpha(s_{j-1}, s_j) + \sum_{j=1,\ j\ \text{even}}^{k} \beta(s_{j-1}, s_j) \right\}, \tag{11}$$

where $\mathbf{z}^*$ is any solution of the problem that minimizes (3) subject to $\mathbf{y} \in Y(k, n)$.

Let $\{s_j : j = 0, 1, \ldots, k\}$ be any integers that satisfy the conditions

$$1 = s_0 \le s_1 \le \cdots \le s_k = n$$

and let $\mathbf{y}^-$ be the n-vector that gives the terms of the expression

$$||\mathbf{y}^- - \boldsymbol{\phi}||_2^2 = \alpha(s_0, s_1) + \sum_{j=2,\ j\ \text{odd}}^{k} \alpha(s_{j-1} + 1, s_j) + \sum_{j=2,\ j\ \text{even}}^{k} \beta(s_{j-1} + 1, s_j), \tag{12}$$

where we define $\alpha(i, j)$ and $\beta(i, j)$ to be zero if $j < i$. Hence, we obtain the inequality

$$\alpha(s_0, s_1) + \sum_{j=2,\ j\ \text{odd}}^{k} \alpha(s_{j-1} + 1, s_j) + \sum_{j=2,\ j\ \text{even}}^{k} \beta(s_{j-1} + 1, s_j) \le$$

$$\sum_{j=1,\ j\ \text{odd}}^{k} \alpha(s_{j-1}, s_j) + \sum_{j=1,\ j\ \text{even}}^{k} \beta(s_{j-1}, s_j). \tag{13}$$

As y^- is in $Y(k, n)$ and z^* is optimal, we have $||z^* - \phi||_2^2 \leq ||y^- - \phi||_2^2$. This inequality, (12) and (13) imply inequality (11). We deduce from (9), (10) and (11) that Equation (7) is true.

In order to complete the proof of the theorem, we consider next the sufficient conditions. We let the integers $\{t_j : j = 0, 1, \ldots, k\}$ satisfy the conditions (2) and the Equation (7) and we show that $y(T)$ minimizes (3) subject to $y \in Y(k, n)$. It suffices to construct a vector with integer variables $\{t_j : j = 0, 1, \ldots, k\}$ that satisfies the constraints satisfied by $y(T)$, minimizes (3) and provides the interpolation equations (4). The remaining proof is rather long and falls into two parts.

In the first part, we construct a vector that gives the least value of (3). As a consequence of the necessary conditions, if the values $\{t_j : j = 1, 2, \ldots, k - 1\}$ are optimal, then the least value of the objective function (3) for y in $Y(k, n)$ is achieved and it is equal to

$$||z^* - \phi||_2 = \sum_{j=1,\ j\ \text{odd}}^{k} \alpha(t_{j-1}, t_j) + \sum_{j=1,\ j\ \text{even}}^{k} \beta(t_{j-1}, t_j), \tag{14}$$

where z^* is any solution of the calculation.

Let ψ be the n-vector whose components occur in the definition of $\alpha(t_{j-1}, t_j - 1)$ when j is odd in $[1, k]$ and in the definition of $\beta(t_{j-1}, t_j - 1)$ when j is even in $[1, k]$. It follows that either $\psi_{t_j-1} \geq \psi_{t_j}$ or $\psi_{t_j-1} < \psi_{t_j}$ when j is odd, where ψ_{t_j-1} occurs at $\alpha(t_{j-1}, t_j - 1)$ and ψ_{t_j} occurs at $\beta(t_j, t_{j+1} - 1)$. By changing t_j if necessary to $t_j - 1$ we can restore the feasibility of the constraints (1). The case when j is even is treated analogously. Therefore $\psi \in Y(k, n)$ and

$$||z^* - \phi||_2 \leq ||\psi - \phi||_2. \tag{15}$$

Since

$$||\psi - \phi||_2^2 = \sum_{j=1,\ j\ \text{odd}}^{k-1} \alpha(t_{j-1}, t_j - 1) + \sum_{j=1,\ j\ \text{even}}^{k-1} \beta(t_{j-1}, t_j - 1) + \delta(t_{k-1}, t_k),$$

$$\tag{16}$$

where δ stands for α if k is odd and for β if k is even, we obtain the bound

$$||\psi - \phi||_2 \leq \sum_{j=1,\ j\ \text{odd}}^{k} \alpha(t_{j-1}, t_j) + \sum_{j=1,\ j\ \text{even}}^{k} \beta(t_{j-1}, t_j). \tag{17}$$

In view of (14), (15), (16) and (17) we obtain the equality

$$\sum_{j=1,\ j\ odd}^{k-1} \alpha(t_{j-1}, t_j - 1) + \sum_{j=1,\ j\ even}^{k-1} \beta(t_{j-1}, t_j - 1) + \delta(t_{k-1}, t_k) =$$

$$\sum_{j=1,\ j\ odd}^{k} \alpha(t_{j-1}, t_j) + \sum_{j=1,\ j\ even}^{k} \beta(t_{j-1}, t_j). \qquad (18)$$

If we remove the condition $y_{t_j-1} \leq y_{t_j}$ from the calculation of $\alpha(t_{j-1}, t_j)$, the minimum value of $\sum_{i=t_{j-1}}^{t_j} (y_i - \phi_i)^2$ subject to $y_{t_{j-1}} \leq y_{t_{j-1}+1} \leq \cdots \leq y_{t_j}$ is not greater than before and it is equal to $\alpha(t_{j-1}, t_j - 1)$. Hence we have the inequalities

$$\alpha(t_{j-1}, t_j - 1) \leq \alpha(t_{j-1}, t_j), \text{ for } j = 1, 2, \ldots, k.$$

Since any strict inequality in these relations will rule out (18), $\alpha(t_{j-1}, t_j - 1) = \alpha(t_{j-1}, t_j)$ should occur, and similarly $\beta(t_{j-1}, t_j - 1) = \beta(t_{j-1}, t_j)$. Therefore, we have obtained the relations

$$\left.\begin{array}{l} \alpha(t_{j-1}, t_j - 1) = \alpha(t_{j-1}, t_j), \ j \ odd \\ \beta(t_{j-1}, t_j - 1) = \beta(t_{j-1}, t_j), \ j \ even \end{array}\right\}, \qquad (19)$$

where, for notational purpose, we let $t_k - 1 = t_k$. This concludes the first part of the proof with respect to the sufficiency conditions.

In the second part, we show that vector $\boldsymbol{\psi}$ satisfies the constraints (1) and the interpolation equations (4). Remembering the monotonicity of the components of $\boldsymbol{\psi}$ on the interval $[t_{j-1}, t_j - 1]$, when $j \in [1, k-1]$ is odd, we let p be an integer such that $t_{j-1} \leq p \leq t_j - 1$ and $\psi_{p-1} < \psi_p = \psi_{p+1} = \cdots = \psi_{t_j-1} = \eta^*$, except that we ignore the inequality $\psi_{p-1} < \psi_p$ if $p = t_{j-1}$, where η^* is the unique number that minimizes the sum of squares $\sum_{i=p}^{t_j-1} (\eta - \phi_i)^2$. Note that η^* is the average of the data $\{\phi_p, \phi_{p+1}, \ldots, \phi_{t_j-1}\}$. Further, we let $\{\psi_i^{(\alpha)} : i = t_{j-1}, t_{j-1} + 1, \ldots, t_j\}$ be the components that occur in the definition of $\alpha(t_{j-1}, t_j)$. In order to calculate $\alpha(t_{j-1}, t_j)$ from $\alpha(t_{j-1}, t_j - 1)$, whenever $t_j - t_{j-1} \geq 1$, we find that $\eta^* > \phi_{t_j}$ would give $\alpha(t_{j-1}, t_j) > \alpha(t_{j-1}, t_j - 1)$, which contradicts the first line of (19). Hence $\eta^* \leq \phi_{t_j}$ and it is optimal to let $\psi_{t_j}^{(\alpha)} = \phi_{t_j}$. Besides that we have derived this equation, the monotonicity condition $\psi_{t_j-1} \leq \psi_{t_j}^{(\alpha)}$ is also satisfied. Hence and since the components $\{\psi_i^{(\alpha)} = \psi_i : i = t_{j-1}, t_{j-1} + 1, \ldots, t_j - 1\}$ are allowed by conditions $y_{t_{j-1}} \leq y_{t_{j-1}+1} \leq \cdots \leq y_{t_j-1}$ and are unique, it follows that the components

$$\psi_i^{(\alpha)} = \begin{cases} \psi_i, & i = t_{j-1}, t_{j-1} + 1, \ldots, t_j - 1 \\ \phi_{t_j}, & i = t_j \end{cases} \qquad (20)$$

must occur in $\alpha(t_{j-1}, t_j)$ while the first line of (19) is being preserved. Thus, we have found that

$$\psi_{t_j-1}^{(\alpha)} \leq \psi_{t_j}^{(\alpha)} = \phi_{t_j}, \ j \text{ odd} \tag{21}$$

and similarly we deduce from the second line of the Equation (19) that

$$\psi_{t_j-1}^{(\beta)} \geq \psi_{t_j}^{(\beta)} = \phi_{t_j}, \ j \text{ even}, \tag{22}$$

where we let $\{\psi_i^{(\beta)} : i = t_{j-1}, t_{j-1} + 1, \ldots, t_j\}$ be the components that occur in the definition of $\beta(t_{j-1}, t_j)$.

Next, we continue the proof by establishing the relations

$$\left. \begin{array}{l} \phi_{t_j} = \psi_{t_j}^{(\beta)} \geq \psi_{t_j+1}^{(\beta)}, \ j \text{ odd} \\ \phi_{t_j} = \psi_{t_j}^{(\alpha)} \leq \psi_{t_j+1}^{(\alpha)}, \ j \text{ even} \end{array} \right\}. \tag{23}$$

To this end, we will first show that

$$\beta(t_j, t_{j+1}) = \beta(t_j + 1, t_{j+1}), \ j \text{ odd}.$$

If we remove the condition $y_{t_j} \geq y_{t_j+1}$ from the calculation of $\beta(t_j, t_{j+1})$, the minimum value of $\sum_{i=t_j}^{t_{j+1}} (y_i - \phi_i)^2$ subject to $y_{t_j+1} \geq y_{t_j+2} \geq \cdots \geq y_{t_{j+1}}$ is not greater than before and it is equal to $\beta(t_j + 1, t_{j+1})$. It follows that we have the inequality $\beta(t_j, t_{j+1}) \geq \beta(t_j + 1, t_{j+1})$. Strict inequality here would imply $\phi_{t_j} < \psi_{t_j}^{(\beta)}$. Hence we let q be an integer such that $t_j < q \leq t_{j+1}$ and $\psi_{t_j}^{(\beta)} = \psi_{t_j+1}^{(\beta)} = \cdots = \psi_q^{(\beta)} > \psi_{q+1}^{(\beta)}$, except that we ignore the inequality $\psi_q^{(\beta)} > \psi_{q+1}^{(\beta)}$ if $q = t_{j+1}$. As $\psi_{t_j}^{(\beta)}$ is equal to the average of the data $\{\phi_{t_j}, \phi_{t_j+1}, \ldots, \phi_q\}$ it follows that either there exists an integer κ such that $t_j < \kappa \leq q$ and $\psi_\kappa^{(\beta)} < \phi_\kappa$ or $\psi_{t_j}^{(\beta)} = \phi_{t_j+1} = \phi_{t_j+2} = \cdots = \phi_q$. We consider these two cases.

(1) If $\psi_\kappa^{(\beta)} < \phi_\kappa$ we can increase $\psi_\kappa^{(\beta)}$ to ϕ_κ, which reduces the value of $||z^* - \phi||_2$ and yet, remembering the monotonicity of $\{\psi_i^{(\alpha)} : i = t_{j-1}, t_{j-1} + 1, \ldots, t_j\}$, the relations (21) and the inequality $\phi_{t_j} < \psi_{t_j}^{(\beta)}$, we obtain the inequalities

$$\psi_{t_j-1}^{(\alpha)} \leq \cdots \leq \psi_{t_j}^{(\alpha)} = \phi_{t_j} < \psi_{t_j}^{(\beta)} = \cdots = \psi_{\kappa-1}^{(\beta)} < \psi_\kappa^{(\beta)} = \phi_\kappa. \tag{24}$$

Hence by changing t_j to the integer κ we can restore the conditions (1), preserve the relations (21) and reduce the optimal value of (3), which is a contradiction.

(2) A similar contradiction is derived if $\psi_{t_j}^{(\beta)} = \phi_{t_j+1} = \phi_{t_j+2} = \cdots = \phi_q$ upon replacing t_j by $t_j + 1$.

We conclude from these two cases that the assumption $\beta(t_j, t_{j+1}) > \beta(t_j + 1, t_{j+1})$ is contradicted and the equation $\beta(t_j, t_{j+1}) = \beta(t_j + 1, t_{j+1})$ follows.

Next, by an argument similar to that given just above (20), it is optimal to let $\psi_{t_j}^{(\beta)} = \phi_{t_j}$ which gives the first line of (23) and similarly we can establish the second line of (23).

We deduce from (21), (22) and (23) that the equations

$$\psi_{t_j}^{(\alpha)} = \phi_{t_j} = \psi_{t_j}^{(\beta)}, \quad j \in [1, k-1] \tag{25}$$

hold. Therefore by letting the components of the vector ψ be

$$\psi_{t_j} = \phi_{t_j}, \quad j = 1, 2, \ldots, k-1 \tag{26}$$

and

$$\psi_i = \begin{cases} \psi_i^{(\alpha)}, & i = t_{j-1}+1, t_{j-1}+2, \ldots, t_j-1, \ j \text{ odd} \\ \psi_i^{(\beta)}, & i = t_{j-1}+1, t_{j-1}+2, \ldots, t_j-1, \ j \text{ even,} \end{cases} \tag{27}$$

in view of (25), (26) and the monotonicity properties of $\psi^{(\alpha)}$ and $\psi^{(\beta)}$ on the intervals $[t_{j-1}, t_j]$, we have the bounds

$$\left. \begin{array}{l} \psi_{t_j-1} \leq \psi_{t_j} \geq \psi_{t_j+1}, \ j \text{ odd} \\ \psi_{t_j-1} \geq \psi_{t_j} \leq \psi_{t_j+1}, \ j \text{ even} \end{array} \right\}, \tag{28}$$

for all integers j in $[1, k-1]$.

Thus, given the integer values $\{t_j : j = 1, 2, \ldots, k-1\}$, the conditions (2) and the Equations (7), we have constructed a vector ψ that satisfies the same constraints as $y(T)$, provides the least value of the function (3) and allows the interpolation equations (26). Since $y(T)$ is the unique solution of the strictly convex quadratic programming problem that minimizes (3) subject to the constraints (1) we have $y(T) = \psi$. The proof of the theorem is complete. □

The important consequence of Theorem 1 is that it reduces the combinatorial problem that defines the least squares piecewise monotonic approximation to the equivalent formulation (7), which partitions the data into disjoint sets of adjacent data and solves a monotonic problem on each set. The partitioning is achieved by dynamic programming (for a general reference on dynamic programming see Bellman [2]) and the procedure of Demetriou and Powell [10] is quite efficient for this calculation. Subsequently we give a brief description.

If the values $\{t_j : j = 0, 1, \ldots, k\}$ are optimal and k is odd, say, then the least value of (3) is the expression

$$\sum_{j=1, \ j \text{ odd}}^{k} \alpha(t_{j-1}, t_j) + \sum_{j=1, \ j \text{ even}}^{k} \beta(t_{j-1}, t_j) =$$

$$\alpha(t_0, t_1) + \beta(t_1, t_2) + \alpha(t_2, t_3) + \cdots + \alpha(t_{k-1}, t_k). \tag{29}$$

Now for any t_2, the value of t_1 from $[t_0, t_2]$ that minimizes $\alpha(t_0, t_1) + \beta(t_1, t_2)$ is independent of the integers $\{t_j : 3 \leq j \leq k\}$. If we let

$$\gamma(2, t_2) = \min_{t_0 \leq t_1 \leq t_2} \{\alpha(t_0, t_1) + \beta(t_1, t_2)\}$$

for $1 \leq t_2 \leq n$, the right-hand side of Equation (29) becomes equal to

$$\gamma(2, t_2) + \alpha(t_2, t_3) + \cdots + \alpha(t_{k-1}, t_k). \tag{30}$$

Proceeding in this way, we obtain

$$\gamma(3, t_3) = \min_{t_0 \leq t_2 \leq t_3} \min_{t_0 \leq t_1 \leq t_2} \{\alpha(t_0, t_1) + \beta(t_1, t_2) + \alpha(t_2, t_3)\}$$

and we generalize by defining

$$\gamma(m, t) = \sum_{i=1}^{t} (z_i - \phi_i)^2,$$

where for any integers $m \in [1, k]$ and $t \in [1, n]$, we let $Y(m, t)$ be the set of t-vectors z with m monotonic sections. Therefore in order to calculate $\gamma(k, n)$, which is the least value of (3), we begin with the values $\gamma(1, t) = \alpha(1, t)$, for $t = 1, 2, \ldots, n$ and proceed by applying the formulae

$$\gamma(m, t) = \begin{cases} \min_{1 \leq s \leq t} [\gamma(m-1, s) + \alpha(s, t)], & m \text{ odd} \\ \min_{1 \leq s \leq t} [\gamma(m-1, s) + \beta(s, t)], & m \text{ even}, \end{cases} \tag{31}$$

and storing $\tau(m, t)$, which is the value of s that minimizes the right-hand term of expression (31), for $t = 1, 2, \ldots, n$, as $m = 2, 3, \ldots, k$. Then $\gamma(k, n)$ can be found in $O(kn^2)$ computer operations. At the end of the calculation, $m = k$ occurs and the value $\tau(k, n)$ is the integer t_{k-1} that is required in equation

$$\gamma(k-1, t_{k-1}) + \alpha(t_{k-1}, n) = \min_{1 \leq s \leq n} [\gamma(k-1, s) + \alpha(s, n)], \quad k \text{ odd}, \tag{32}$$

if k is odd and analogously if k is even. Hence, we set $t_0 = 1$ and $t_k = n$, and we obtain the sequence of optimal values $\{t_j : j = 1, 2, \ldots, k-1\}$ by the backward formula

$$t_{j-1} = \tau(j, t_j), \text{ for } j = k, k-1, \ldots, 2. \tag{33}$$

Finally, the components of an optimal approximation are obtained by independent monotonic approximation calculations between adjacent $\{t_j\}$.

Further considerations on improved versions of this calculation are given in [6] and the relevant Fortran software package [7], which is available through the Collected Algorithms in the Transactions on Mathematical Software (http://calgo.acm.org/). Recent work of the author [9], which takes advantage of some ordering relations of the integers $\tau(m, t)$, implements the dynamic programming formulae (31) in terms of a tree structure, which reduces the computational complexity to $O(n^2 + kn \log_2 n)$. Both algorithms give the same solution, but it is the utilization of an algorithm that may assist the choice.

The mentioned calculations include also techniques that reduce operations count in theory and achieve far greater reductions of the computation times in practice. With no further details, we quote some timings from such algorithms, when they calculate an approximation as follows. We let the number of data take the values $n = 100, 1000, 5000, 10{,}000, 20{,}000$ and $30{,}000$, we let ϵ_i be a random number from the distribution that is uniform on the interval $[-0.1, 0.1]$ and we generated the data $\phi_i = \sin(\pi x_i) + \epsilon_i$ on the equally spaced grids $0 = x_1 < x_2 < \cdots < x_n = u$, for $u = 9, 24, 49$. We required best approximations with $k = 10, 25, 50$ monotonic sections on the first, second and third grid, respectively. The experiments were run on a HP 8770w portable workstation with an Intel Core i7-3610QM, 2.3 GHz processor, which was used with the standard Fortran compiler of the Intel Visual Fortran Composer XE2013 in single precision arithmetic operating on Windows 7 with 64 bits word length. We present the required CPU times in seconds in Tables 1 and 2 by using versions from two algorithms. The times of Table 1 seem to be weakly proportional to n^2 and the times of Table 2 proportional to n. Recalling the number of combinations required in order to obtain the optimal integers t_j, $j = 1, 2, \ldots, k - 1$, we see that, indeed, the algorithms are very efficient.

Table 1 Tabulation of CPU time to apply the software package of [7]

n	$k = 10$	$k = 25$	$k = 50$
100	0.02	0.01	0.00
1000	0.05	0.03	0.03
5000	0.50	0.66	0.53
10,000	2.09	1.73	1.72
20,000	4.93	7.72	10.65
30,000	3.81	15.79	25.18

Table 2 As in Table 1, but the $O(n^2 + kn \log_2 n)$ algorithm is applied

n	$k = 10$	$k = 25$	$k = 50$
100	0.00	0.00	0.01
1000	0.00	0.02	0.01
5000	0.02	0.06	0.09
10,000	0.03	0.14	0.23
20,000	0.14	0.30	0.55
30,000	0.23	0.55	0.95

3 Estimation of Peaks of an NMR Spectrum

In this section we present an example of our method intended to illustrate the estimation of peaks in an NMR spectrum sample from chemical shifts of CE-584.1, which is the morphoagronomic characteristic of a cowpea seed (vigna unguiculata). In general, the location of peaks and their intensities for a spectrum are the signature of a sample. The complexity of the underlying physical laws makes this a good test of the power of the piecewise monotonic approximation method in peak finding.

To be specific, we downloaded the datafile of a spectrum, which accompanies the article of Alves Filho et al. [1] and is available on the Elsevier website [12]. The spectrum was sampled on an NMR spectrometer Agilent 600-MHz, 5mmH (H-F/^{15}N-^{31}P) One Probe^{TM}. The raw data is provided in file 'Complimentary Table 3.xls' of [1]. The leftist two columns, where the first column keeps the chemical shift (ppm) and the second column (CE584.1) keeps the intensity, provide the values $\{x_i : i = 1, 2, \ldots, n\}$ and $\{\phi_i : i = 1, 2, \ldots, n\}$, respectively for our calculation. The abscissae $\{x_i : i = 1, 2, \ldots, n\}$ are irrelevant to our calculation, except that they are used in our plots. The file contains 110,403 pairs of data far too many to be presented as raw numbers in these pages. Although the details cannot be seen due to page resolution, we may capture the main features of this data set by looking at Figure 1. Indeed, we can see, for instance, very small deviations, many distinguishable peaks,

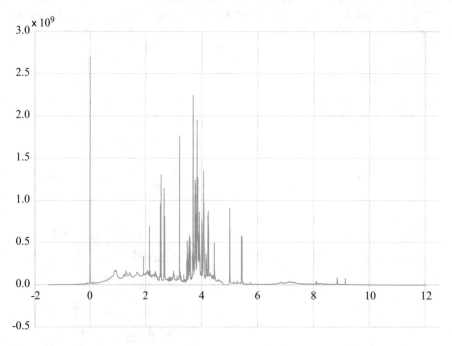

Fig. 1 ^{1}H NMR spectrum dataset of Vigna unguiculata 584.1 (See Table 3 in [1]). The solid line joins 110,403 data points

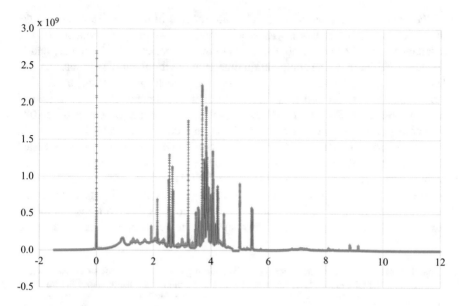

Fig. 2 Best piecewise monotonic fit with $k = 50$ to the data (crosses) of Figure 1. The solid line illustrates the fit

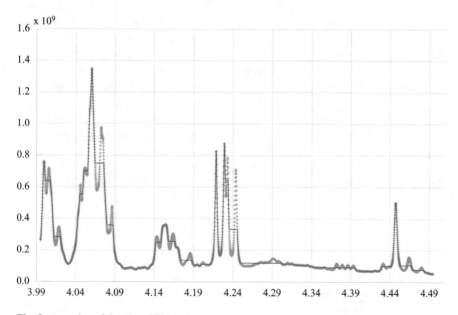

Fig. 3 A portion of the plot of Figure 2

sharp increases and several peaks with lower intensity. However, it is not possible to delve into details.

We seek major turning points (peaks and troughs) of this spectrum. We fed the data to our computer program with $k = 50$ without any preliminary analysis. The resultant fit has 49 turning points, so there are 25 peaks and 24 troughs, and is displayed in Figure 2. The computation time required for obtaining the solution was less than a minute. We see that the fit to the data is much smoother than are the data values themselves, but still it is difficult to distinguish many details on a printed page. Therefore, in Figure 3 we display a portion of Figure 2, which includes 4089 data points or about 4% of the amount of data, in order to see local details of the fit.

It is important to note that the piecewise monotonic approximation has revealed the most important turning points (peaks and troughs), while it interpolates the data at these points due to the Equation (4). In view of this property of the approximation, we present in Table 3 the positions of the turning points of the fit together with the corresponding intensities. Two points are worth emphasizing with respect to this example. Firstly, the method effectively captured the trends of the data and detected

Table 3 Positions (x_{t_j}) and intensities (ϕ_{t_j}) of the turning points in the NMR spectrum of Vigna unguiculata 584.1 (see Figure 1) by the best fit with $k = 50$ monotonic sections (see Figure 2)

j	t_j	x_{t_j}	ϕ_{t_j}	j	t_j	x_{t_j}	ϕ_{t_j}
0	1	−1.499966e+0	−3.090880e+5	26	43440	3.811764e+0	1.837607e+8
1	12268	4.540000e−5	2.700000e+9	27	43566	3.827171e+0	1.952043e+9
2	12914	7.903843e−2	1.709395e+7	28	43742	3.848692e+0	5.128544e+8
3	19909	9.343884e−1	1.734959e+8	29	43818	3.857985e+0	1.268841e+9
4	21242	1.097388e+0	7.505543e+7	30	44068	3.888556e+0	2.796269e+8
5	29753	2.138115e+0	6.962561e+8	31	44259	3.911911e+0	8.618427e+8
7	32950	2.529044e+0	9.596586e+8	33	44974	3.999341e+0	7.619820e+8
8	33049	2.541150e+0	9.292265e+7	34	45230	4.030645e+0	1.102941e+8
9	33158	2.554479e+0	1.304342e+9	35	45472	4.060237e+0	1.349148e+9
10	33672	2.617331e+0	5.020448e+7	36	45922	4.115263e+0	7.794741e+7
11	34024	2.660373e+0	1.140659e+9	37	46246	4.154882e+0	3.639329e+8
12	34146	2.675292e+0	1.006428e+8	38	46598	4.197925e+0	9.973205e+7
13	34232	2.685808e+0	8.102999e+8	39	46765	4.218345e+0	8.292408e+8
14	35320	2.818848e+0	4.478114e+7	40	46808	4.223603e+0	1.653756e+8
15	38513	3.209289e+0	1.762919e+9	41	46851	4.228861e+0	8.764532e+8
16	39442	3.322887e+0	3.736386e+7	42	48387	4.416684e+0	6.837490e+7
17	40733	3.480751e+0	5.216682e+8	43	48634	4.446887e+0	5.032446e+8
18	40935	3.505452e+0	6.344673e+7	44	51113	4.750020e+0	0.000000e+0
19	41487	3.572950e+0	5.599327e+8	45	53194	5.004485e+0	9.085181e+8
20	41857	3.618194e+0	6.247412e+7	46	54734	5.192796e+0	1.524049e+7
21	42414	3.686304e+0	2.244380e+9	47	56613	5.422561e+0	5.838891e+8
22	42678	3.718586e+0	1.204448e+8	48	56664	5.428797e+0	7.275355e+7
23	43036	3.762362e+0	1.240367e+9	49	56726	5.436378e+0	5.660183e+8
24	43094	3.769455e+0	4.398070e+8	50	110403	1.200001e+1	−4.437627e+5
25	43133	3.774224e+0	1.202403e+9	–	–	–	–

appropriate peaks subject to the requirement that $k = 50$ in the constraints (1) and the conditions (2). Secondly, the method is so efficient that we can run it for a sequence of integers k if a suitable value is not known in advance.

4 Summary

We considered the problem that makes least the sum of squares of residuals by an approximation to n noisy data subject to the condition that the approximated values have at most k monotonic sections, where the joints of the monotonic sections are also unknowns of the calculation. It is a combinatorial optimization calculation that can have an enormous number of local minima.

We stated a characterization theorem, which provides a decomposition of the problem into at most k best monotonic approximation problems to disjoint subsets of adjacent data. Decomposition is achieved by a dynamic programming procedure, which allows some highly efficient calculations of the solution. We quoted some timings from numerical experiments which confirmed these efficiencies.

Piecewise monotonic approximation may have many useful applications in various fields. For example, one may require to calculate the peaks of a spectrum, when the spectrum is represented by a number of noisy measurements, which is a problem at the heart of spectroscopy and chromatography. Also, medical applications of this method arise from image restoration in magnetic resonance imaging and computed tomography, such as in the treatment of malignant tumors by radiation, when dose reduction techniques reduce the diagnostic quality of the image due to noise contamination. We demonstrated the efficacy of our method on peak finding of a large dataset of NMR data and some further results for this case will be published separately.

It is straightforward to generalize Theorem 1 to the case when one requires a best piecewise monotonic approximation that minimizes an error function that instead of $(y_i - \phi_i)^2$ the function $h_i(y_i - \phi_i)$ is considered, where h_i is a strictly convex and continuous function from $\mathbb{R}$ to $\mathbb{R}$ whose least value is $h_i(0) = 0$ and $h_i(\theta) \to \infty$ as $|\theta| \to \infty$.

Acknowledgement I want to express my deep gratitude to late Professor M.J.D. Powell of Cambridge University for his invaluable advice, guidance and discussions on this problem since when I was his PhD student [4].

References

1. Alves Filho, E.G., Silva, L.M.A., Teofilo, E.M., Larsen, F.H., de Brito, E.S.: ^{1}H NMR spectra dataset and solid-state NMR data of cowpea (Vigna unguiculata). Data Brief **11**, 135–146 (2017)
2. Bellman, R.E.: Dynamic Programming. Princeton University Press, Princeton (1957)

3. Cullinan, M.P., Powell, M.J.D.: Data smoothing by divided differences. In: Watson, G.A. (ed.) Numerical Analysis Proceedings, Dundee 1981. Lecture Notes in Mathematics, vol. 912, pp. 26–37. Springer, Berlin (1982)
4. Demetriou, I.C.: Data Smoothing by Piecewise Monotonic Divided Differences, Ph.D. Dissertation, Department of Applied Mathematics and Theoretical Physics, University of Cambridge, Cambridge (1985)
5. Demetriou, I.C.: Best L_1 piecewise monotonic data modelling. Int. Trans. Oper. Res. **1**(1), 85–94 (1994)
6. Demetriou, I.C.: Discrete piecewise monotonic approximation by a strictly convex distance function. Math. Comput. **64**, 157–180 (1995)
7. Demetriou, I.C.: Algorithm 863: L2WPMA, a Fortran 77 package for weighted least-squares piecewise monotonic data approximation. ACM Trans. Math. Softw. **33**(1), 1–19 (2007)
8. Demetriou, I.C.: A characterization theorem for the best L_1 piecewise monotonic data approximation problem. In: Pardalos, P.M., Rassias, T.M. (eds.) Contributions in Mathematics and Engineering. In Honor of Constantin Caratheodory (Foreword by R. Tyrrell Rockafellar), pp. 117–126. Springer, Switzerland (2016)
9. Demetriou, I.C.: A binary search algorithm for best piecewise monotonic data approximation, 37 pp. Report Department of Economic 2019/1, National and Kapodistrian University of Athens (2019)
10. Demetriou, I.C., Powell, M.J.D.: Least squares smoothing of univariate data to achieve piecewise monotonicity. IMA J. Numer. Anal. **11**, 411–432 (1991)
11. Fletcher, R.: Practical Methods of Optimization. Wiley, Chichester (2003)
12. https://www.elsevier.com/journals/data-in-brief/2352-3409?generatepdf=true. Accessed on 4 Aug 2018
13. van Eeden, C.: Maximum likelihood estimation of ordered probabilities. Indag. Math. **18**, 444–455 (1956)

Recent Progress in Optimization of Multiband Electrical Filters

Andrei Bogatyrëv

Abstract The best uniform rational approximation of the *sign* function on two intervals was explicitly found by Russian mathematician E.I. Zolotarëv in 1877. The progress in math eventually led to the progress in technology: half a century later German electrical engineer and physicist W. Cauer on the basis of this solution has invented low- and high-pass electrical filters known today as elliptic or Cauer-Zolotarëv filters and possessing the unbeatable quality. We discuss a recently developed approach for the solution of optimization problem naturally arising in the synthesis of multi-band (analogue, digital or microwave) electrical filters. The approach is based on techniques from algebraic geometry and generalizes the effective representation of Zolotarëv fraction.

1 History and Background

Sometimes the progress in mathematics brings us to the progress in technology. One of such examples is the invention of low- and high-pass electrical filters widely used nowadays is electronic appliances. The story started in year 1877 when E.I. Zolotarëv (1847–1878)—the pupil of P.L. Chebyshëv—has solved a problem of best uniform rational approximation of the function $sign(x)$ on two segments of real axis separated by zero. His solution now called *Zolotarëv fraction* is the analogy of Chebyshëv polynomials in the realm of rational functions and inherits many nice properties of the latter. This work of Zolotarëv who also attended lectures of K. Weierstrass and corresponded to him was highly appreciated by the German scholar. More than 50 years later German electrical engineer, physicist and guru of network synthesis Wilhelm Cauer (1900–1945) has invented electrical filters with the transfer function based on Zolotarëv fraction.

A. Bogatyrëv (✉)
Institute for Numerical Mathematics, Russian Academy of Sciences, Moscow, Russia

© Springer Nature Switzerland AG 2019
I. C. Demetriou, P. M. Pardalos (eds.), *Approximation and Optimization*,
Springer Optimization and Its Applications 145,
https://doi.org/10.1007/978-3-030-12767-1_8

Further development of technologies brings us to more sophisticated optimization problems [13, 14, 20, 21, 23]. In particular, modern gadgets may use several standards of wireless communication like IEEE 806.16, GSM, LTE, GPS and therefore a problem of filtering on several frequency bands arises. Roughly, the problem is this: given the mask of a filter, that is the boundaries of its stop and pass bands, the levels of attenuation at the stopbands and the permissible ripple magnitude at the passbands, to find minimum degree real rational function fitting this mask. The problem reduces to a solution of a series of somewhat more simple minimal deviation problems on several segments similar to the one considered by Zolotarëv. Several equivalent formulations will appear in Section 2.1.

Those problems turned out to be very difficult from the practical viewpoint because of intrinsic instability of most numerical methods of rational approximation. However, we know how the 'certificate' of the solution (see contribution from Panos Pardalos in this volume) for this particular case looks like: the solution possesses the so-called *equiripple property*, that is behaves like a wave of constant amplitude on each stop or passband. The total number of ripples is bounded from below. In a sense the solution for this problem is rather simple—you just manifest function with a suitable equiripple property. Such behaviour is very unusual for generic rational functions; therefore, functions with equiripple property fill in a variety of relatively small dimension in the set of rational functions of bounded degree. The natural idea is to look for the solution in the 'small' set of the distinguished functions instead of the 'large' set of generic functions. *Ansatz* is an explicit formula with few parameters which allows to parametrize the 'small' set. This Ansatz ideology had been already used to calculate the so-called Chebyshëv polynomials on several segments [4], optimal stability polynomials for explicit multistage *Runge-Kutta* methods [5, 6] and solve some other problems [7]. Recall, e.g., *Bethe Ansatz* for finding exact solutions for Heisenberg antiferromagnetic model. Ansatz for optimal electrical filters is discussed in Section 7.

2 Optimization Problem for Multiband Filter

Suppose we have a finite collection E of disjoint closed segments of real axis $\mathbb{R}$. The set has a meaning of frequency bands and is decomposed into two subsets: $E = E^+ \cup E^-$ which are, respectively, called the passbands E^+ and the stopbands E^-. Both subsets are nonempty. Optimization problem for electrical filter has several equivalent settings [1, 3, 12, 22, 26].

2.1 Four Settings

In each of the listed below cases we minimize certain quantity among real rational functions $R(x)$ of bounded degree $\deg R \leq n$ being the maximum of the degrees of numerator and denominator of the fraction. The goal function may be the following.

2.1.1   Minimal Deviation

$$\frac{\max_{x \in E^+} |R(x)|}{\min_{x \in E^-} |R(x)|} \longrightarrow \min =: \theta^2 \le 1$$

2.1.2   Minimal Modified Deviation

$$\max\{\max_{x \in E^+} |R(x)|, \max_{x \in E^-} 1/|R(x)|\} \longrightarrow \min =: \theta$$

2.1.3   Third Zolotarëv Problem

Minimize θ under the condition that there exist real rational function $R(x)$, $\deg R \le n$, with the restrictions

$$\min_{x \in E_-} |R(x)| \ge \theta^{-1}, \qquad \max_{x \in E_+} |R(x)| \le \theta$$

2.1.4   Fourth Zolotarëv Problem

Define the indicator function $S(x) = \pm 1$ when $x \in E^{\pm}$. Find the best uniform rational approximation $R(x)$ of $S(x)$ of the given degree:

$$||R - S||_{C(E)} := \max_{x \in E} |R(x) - S(x)| \to \min =: \mu.$$

It is a good exercise to show that all four settings are equivalent and in particular the value of θ is the same for the first three settings and $1/\mu = (\theta + 1/\theta)/2$ for the fourth one.

2.2   Study of Optimization Problem

Setting 2.1.1 appears in the paper [3] by R.A.-R. Amer, H.R. Schwarz (1964). It was transformed to problem of Section 2.1.2 by V.N. Malozemov [22]. Setting 2.1.3 appears after suitable normalization of the rational function in Section 2.1.1 and essentially coincides with the third Zolotarëv problem [26]. Setting 2.1.4 coincides with the fourth Zolotarëv problem [26] and was studied by N.I. Akhiezer [1]. The latter noticed that already in the classical Zolotarëv case when the set E contains just two components, the minimizing function is not unique. This phenomenon was fully explained in the dissertation of R.-A.R. Amer [3] who decomposed the space of rational functions of bounded deviation (defined in the left-hand side of formula in Section 2.1.1) into classes. Namely, it is possible to fix the sign of the polynomial in the numerator of the fraction on each stopband and fix the sign of denominator polynomial on each passband. Then in the closure of each nonempty class there is a unique minimum. All mentioned authors established that (local) minimizing functions are characterized by *alternation* (or *equiripple* in terms of electrical engineers) property. For instance, in the fourth Zolotarëv problem the approximation error $\delta(x) := R(x) - S(x)$ of degree n minimizer has $2n + 2$ *alternation* points $a_s \in E$ where $\delta(a_s) = \pm||\delta||_{C(E)}$ with consecutive change of sign.

3   Zolotarëv Fraction

E.I. Zolotarëv has solved the problem 2.1.4 for the simplest case: $E^{\pm} = \pm[1, 1/k]$, $0 < k < 1$ when $S(x) = sign(x)$. His solution is given parametrically in terms of elliptic functions and its graph (distorted by a pre- composition with a linear fractional map) is shown in Figure 1.

To give an explicit representation for this rational function, we consider a rectangle of size $2 \times |\tau|$:

$$\Pi_\tau = \{u \in \mathbb{C} : \quad |Re\, u| \le 1, 0 < Im\, u < |\tau|\}, \qquad \tau \in i\mathbb{R}^+$$

which may be conformally mapped to the upper half plane with the normalization $x_\tau(u) : \Pi_\tau, -1, 0, 1 \longrightarrow \mathbb{H}, -1, 0, 1$. The latter mapping has a closed appearance $x_\tau(u) = sn(K(\tau)u|\tau)$ in terms of *elliptic sine* and *complete elliptic integral* $K(\tau) = \frac{\pi}{2}\theta_3^2(\tau)$, both of modulus τ [2]. Zolotarëv fraction has a parametric representation resembling the definition of a classical Chebyshëv polynomial:

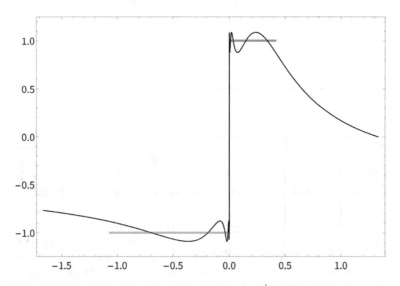

Fig. 1 Graph of Zolotarëv fraction adapted to two segments $E^{\pm}$ of different lengths

Fig. 2 Large rectangle $\Pi_{n\tau}$ composed of n copies of small one Π_τ

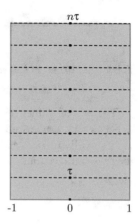

$$Z_n(x_{n\tau}(u)) = x_\tau(u).$$

Of course, it takes some effort to prove that Z_n is the rational function of its argument (we face the same difficulty with classical Chebyshev polynomial defined parametrically as $T_n(\cos(u)) := \cos(nu)$). The qualitative graph of Zolotarëv fraction completely follows from Figure 2, for instance its $2n - 2$ critical points correspond to the interior intersection points of the vertical boundaries of the large rectangle $\Pi_{n\tau}$ and horizontal boundaries of smaller rectangles. Alternation points different from critical points of the fraction correspond to four corners of the large rectangle. Zeros/poles of the fraction correspond to $u = l\tau$ with even/odd l.

Remark 1 Zolotarëv fractions share many interesting properties with Chebyshëv polynomials as the latter are the special limit case of the former [9, 10]. For instance, the superposition of suitably chosen Zolotarëv fractions is again a Zolotarëv fraction. They also appear as the solutions to many other extremal problems [15, 16, 18].

4 Projective View

Here we discuss the optimization problem setting which embraces all the formulations we met before in Section 2.1. We do not treat the infinity point both in the domain of definition and the range of rational function as exceptional. Real line extended by a point at infinity becomes a *real projective line* $\mathbb{R}P^1 := \hat{\mathbb{R}} = \mathbb{R} \cup \{\infty\}$ which is a topological circle. We consider two collections of disjoint closed segments on the extended real line: E consisting of $m \geq 2$ segments and F of just two segments. The segments of both E and F are of two types: $E := E^+ \cup E^-$; $F := F^+ \cup F^-$.

Definition 1 We introduce the class $\mathcal{R}_n(E, F)$ of real rational functions $R(x)$ of a fixed degree $\deg R = n$ such that $R(E^+) \subset F^+$ and $R(E^-) \subset F^-$.

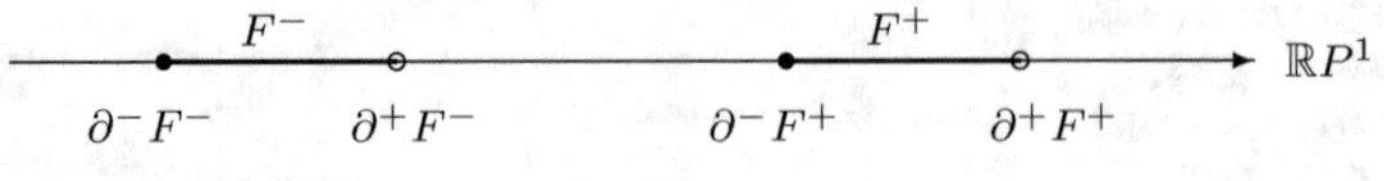

Fig. 3 The ordering of four endpoints ∂F and their colors

The set of values F modulo *projective (=linear-fractional) transformations* depends on a single value—cross ratio of its endpoints. Suppose the endpoints ∂F are cyclically ordered as follows: $\partial^- F^-$, $\partial^+ F^-$, $\partial^- F^+$, $\partial^+ F^+$—see Figure 3 then the cross ratio of four endpoints we define as follows:

Definition 2

$$\kappa(F) := \frac{\partial^+ F^+ - \partial^+ F^-}{\partial^+ F^+ - \partial^- F^+} : \frac{\partial^- F^- - \partial^+ F^-}{\partial^- F^- - \partial^- F^+} > 1.$$

The classes $\mathcal{R}_n(E, F)$ (possibly empty) and the value of the cross ratio have several easily checked properties:

Lemma 1

1. Monotonicity.

$$\mathcal{R}_n(E, F') \subset \mathcal{R}_n(E, F) \qquad \text{once } F' \subset F.$$

2. Projective invariance. For any projective transformations $\alpha, \beta \in PGL_2(\mathbb{R})$,

$$\mathcal{R}_n(\alpha E, \beta F) = \beta \circ \mathcal{R}_n(E, F) \circ \alpha^{-1}.$$

3. The value $\kappa(F)$ is decreasing with the growth of its argument: if $F' \subset F$ then $\kappa(F') > \kappa(F)$.

4.1 Projective Problem Setting

Fix degree $n > 0$ and set $E \subset \mathbb{R}P^1$, find

$$\varkappa(n, E) := \inf\{\kappa(F) : \qquad \mathcal{R}_n(E, F) = \emptyset\}.$$

The idea behind this optimization is the following: we squeeze the set of values F, the functional class $\mathcal{R}_n(E, F)$ diminishes and we have to catch the moment— quantitatively described by the cross ratio $\kappa(F)$—when the class disappears.

Remark 2

1) In problem formulation 2.1.3 the set $F^+ = [-\theta, \theta]$ and the set $F^- = [1/\theta, -1/\theta]$; $\kappa(F) = \left(\frac{1}{2}(\theta + 1/\theta)\right)^2$. In setting 2.1.4 the sets $F^\pm = \pm[1 - \mu, 1 + \mu]$ and $\kappa(F) = \mu^{-2}$.
2) Notice that the cross ratio depends on the order of four participating endpoints and may take six values interchanged by the elements of the so-called unharmonic group.

4.2   Decomposition into Subclasses

Now we decompose each set $\mathcal{R}_n(E, F)$ into subclasses which were first introduced for the problem setting 2.1.1 by R.A.-R. Amer in his PhD thesis [3] in 1964. The construction of these subclasses is purely topological: suppose we identify opposite points of a circle S^1, we get a double cover of a circle identified with real projective line $\mathbb{R}P^1$ by another circle S^1. Now we try to lift the mapping $R(x) : \mathbb{R}P^1 \to \mathbb{R}P^1$ to the double cover of the target space: $\tilde{R}(x) : \mathbb{R}P^1 \to S^1$. There is a topological obstruction to the existence of $\tilde{R}$: the mapping degree or the winding number of $R(x)$ modulo 2. A simple calculation shows that this value is equal to algebraic degree $\deg R \ mod \ 2$. However this lift exists on any simply connected piece of $\mathbb{R}P^1$. Suppose the segment $E(j)$ is made up of two consecutive segments E_j, E_{j+1} of the set of bands $E = \cup_{j=1}^m E_j$ and the gap between them. The set $F \subset \mathbb{R}P^1$ lifted to the circle S^1 consists of four components cyclically ordered as $F_0^-, F_0^+, F_1^-, F_1^+ \subset S^1$. The mapping $R(x) : E(j) \to \mathbb{R}P^1$ has two lifts to the covering circle S^1 and exactly one of them has values $\tilde{R}(x) \in F_0 := F_0^- \cup F_0^+$ when $x \in E_j$. On the opposite side E_{j+1} of the segment the same function $\tilde{R}(x)$ takes values in the set $F_{\sigma(j)}$ with well-defined $\sigma(j) \in \{0, 1\}$. Totally, the function $R(x)$ defines an element of $\mathbb{Z}_2$ for any two consecutive segments of the set E with the only constraint

$$\sum_{j=1}^m \sigma(j) = \deg R \ mod \ 2.$$

which defines the element $\Sigma := (\sigma_1, \sigma_2, \ldots, \sigma_{m-1}) \in \mathbb{Z}_2^{m-1}$. All elements $R(x) \in \mathcal{R}_n(E, F)$ with the same value of $\Sigma \in \mathbb{Z}_2^{m-1}$ make up a subclass $\mathcal{R}_n(E, F, \Sigma)$. Again, one readily checks the properties of the new classes:

Lemma 2

1. Monotonicity:

$$\mathcal{R}_n(E, F', \Sigma) \subset \mathcal{R}_n(E, F, \Sigma) \qquad once \ F' \subset F.$$

2. Projective invariance

$$\beta \circ \mathcal{R}_n(E, F, \Sigma) = \mathcal{R}_n(E, \beta F, \beta \Sigma), \qquad \beta \in PGL_2(\mathbb{R}).$$

here projective transformation β acts on Σ component wise: $\sigma(j)$ reverses exactly when β changes the orientation of projective line and the bands E_j, E_{j+1} are of different $\pm$-type. Otherwise—iff $\beta \in PSL_2(\mathbb{R})$ or bands E_j, E_{j+1} are both pass- or stopbands—$\sigma(j)$ is kept intact.

Remark 3 R.-A.R. Amer [3] combines classes $\mathcal{R}_n(E, F, \Sigma)$ and $\mathcal{R}_n(E, F, \beta \Sigma)$ for β reversing the orientation of projective line and conserving the components $F^{\pm}$. This is why he gets twice less number 2^{m-2} of classes.

4.3 Extremal Problem for Classes

Given degree n, set of bands E, and the class Σ—find

$$\varkappa(n, E, \Sigma) := \inf\{\kappa(F) : \qquad \mathcal{R}_n(E, F, \Sigma) = \emptyset\}. \tag{1}$$

4.4 Equiripple Property

Definition 3 We say that cyclically ordered (on projective line) points $a_1, a_2, \ldots a_s \subset E$ make up an alternation set for the function $R(x) \in \mathcal{R}_n(E, F)$ iff $R(x)$ maps each of those points to the boundary $\partial F = \partial^+ F \cup \partial^- F$, and any two consecutive points—to different sets $\partial^+ F$, $\partial^- F$ colored black and white in Figure 3.

Theorem 1 *If the value $\varkappa(n, E, \Sigma) > 1$, then the closure of the extremal class $\mathcal{R}_n(E, F, \Sigma)$ contains a unique function $R(x)$ which is characterized by the property of having at least $2n + 2$ alternation points when $R(x)$ is not at the boundary of the class.*

Proof of this theorem and other statements of the current section will be given elsewhere.

5 Problem Genesis: Signal Processing

There are many parallels between analogue and digital electronics, this is why many engineering solutions of the past have moved to the new digital era. In particular, the same optimization problem for rational functions discussed in Section 2.1 arises in the synthesis of both analogue and digital electronic devices.

From the mathematical viewpoint electronic device is merely a linear operator which transforms input signals $x(\cdot)$ to output signals $y(\cdot)$. By signals they mean functions of one continuous or discrete argument: $x(t)$, $t \in \mathbb{R}$ or $x(k)$, $k \in \mathbb{Z}$. For technical simplicity they assume that signals vanish in the 'far past'. Another natural assumption that a device processing a delayed signal gives the same but (equally) delayed output which mathematically means that operator commutes with the time shifts. As a consequence, the operator consists in a (discrete) convolution of the input signal with the certain fixed signal—the response $h(\cdot)$ to (discrete) delta function input:

$$y(t) = \int_{\mathbb{R}} h(t')x(t - t')dt'; \qquad y(k) = \sum_{k' \in \mathbb{Z}} h(k')x(k - k').$$

The causality property means that the output cannot appear before the input and implies that *impulse response* $h(\cdot)$ vanishes for negative arguments. Further restrictions on the *impulse response* follow from the physical construction of the device.

Analogue device is an electric scheme assembled of elements like resistors, capacitors, (mutual) inductances, etc. which is governed by Kirchhoff laws. Digital device is governed by the recurrence relation:

$$y(m) := \sum_{j=0}^{n} p_j x(m - j) + \sum_{j=1}^{n} q_j y(m - j), \qquad m \in \mathbb{Z}. \qquad (2)$$

To compute the impulse response, we use the Fourier transform of continuous signals and Z-transform of digital ones (here we do not discuss any convergence):

$$\hat{x}(\omega) := \int_{\mathbb{R}} x(t) \exp(i\omega t)dt; \quad \omega \in \mathbb{H} \qquad \hat{x}(z) := \sum_{k \in \mathbb{Z}} x(k)z^k, \quad z \in \mathbb{C}. \qquad (3)$$

Using the explicit relation (2) for digital device and its Kirchhoff counterpart for analogue ones we observe that the images of input signals are merely multiplied by *rational functions* $\hat{h}(\cdot)$ of appropriate argument. Since the impulse response is real valued, its image—also called the *transfer function*—has the symmetry

$$\hat{h}(-\bar{\omega}) = \overline{\hat{h}(\omega)}, \qquad \hat{h}(\bar{z}) = \overline{\hat{h}(z)}, \qquad \omega, z \in \mathbb{C}.$$

In practice we can physically observe the absolute value of the transfer function: if we 'switch on' a harmonic signal of a given frequency as the input one, then after certain transition process the output signal will also become harmonic, however with a different amplitude and phase. The magnification of the amplitude as a function of frequency is called the magnitude response function and it is exactly equal to the absolute value of transfer function of the device.

Multiband filtering consists in constructing a device which almost keeps the magnitude of a harmonic signal with the frequency in the passbands and almost eliminates signals with the frequency in the stopbands. We use the word 'almost' since the square of the magnitude response is a rational function on the real line (for analogue case):

$$|\hat{h}(\omega)|^2 = \hat{h}(\omega)\hat{h}(-\omega) = R(\omega^2), \qquad \omega \in \mathbb{R}. \tag{4}$$

At best we can talk of approximation of an indicator function which is equal 0 at the stopbands and 1 at the passbands. For certain reasons discussed, e.g., by W. Cauer, uniform (or Chebyshev) approximation is preferable for this practical problem. So we immediately arrive at the fourth Zolotarev problem taking the square of frequency as a new variable. For the digital case we get a similar problem set on the segments of the unit circle which can be transformed to the problem on a real line.

Note that the reconstruction of the *transfer function* $\hat{h}(\cdot)$ from the *magnitude response* is not unique: we have to solve Equation (4) given its right-hand side, which has some freedom. This freedom is used to meet another important restriction on the image of impulse response which is prescribed by the causality: $\hat{h}(\cdot)$ can only have poles in the lower half plane $-\mathbb{H}$ of complex variable ω for the analogue case or strictly outside the unit circle of variable z for the digital case.

Minimal deviation problem in any of the given above settings is just an intermediate step to the following problem of great practical importance. *Find minimal degree filter meeting given filter specifications* like the boundaries of the pass- and stopbands attenuation at the stopbands and allowable ripple amplitude at the passbands. The degree of the rational function $\hat{h}$ is directly related to the complexity of device structure, its size, weight, cost of production, energy consumption, cooling, etc.

6 Approaches to Optimization

There are several major approaches for the practical solution of optimization problem of multiband electrical filter. Three of those are discussed below. Along with them we would single out a computationally efficient Caratheodory-Fejer method (known also as AAK-approximation) solving the problem in the class of quasirational functions which may be further truncated to rational at a cost of extra error [19].

6.1 Remez-Type Methods

Direct numerical optimization is usually based on Remez-type methods. This is a group of algorithms specially designed for uniform rational approximation

[17, 21, 24, 25]. They iteratively build the necessary alternation set for the error function of approximation. Unfortunately the intrinsic instability of Remez algorithms does not allow to get high degree solutions and therefore sophisticated filter specifications. For instance, standard double precision accuracy 10^{-15} used, e.g., in MATLAB does not allow to get solutions of degree n greater than 15–20. We know an example when approximation of degree $n \approx 2000$ required mantissa of 150,000 decimal signs for stability of intermediate computations. Writing just one number of this precision requires 75 standard pages—this is the volume of a typical PhD thesis. Another problem of this group of algorithms is the choice of initial approximation. The set of suitable starting points may have infinitesimal volume.

6.2   Composite Filters

Practical approach of engineers is to decompose complicated problem into many simple ones and solve them one by one. In case of filter synthesis they use a battery of single passband (say, Cauer) filters. This approach is very reliable: it always gives working solutions which however are far from being optimal. We get a substantial rise in the order of filter, and therefore complexity of its structure and the downgrading of many consumer properties.

6.3   Ansatz Method

Is based on an explicit analytical formula for the solution generalizing formula for Zolotarëv fractions. However this formula contains unknown parameters, both continuous and discrete which have to be evaluated given the input data of the problem. Of all approaches this one is the least studied from the algorithmic viewpoint and its usage is restrained by involved mathematical apparatus [8]. Nonetheless it copes with very involved filter specifications: narrow transition bands, large number of working bands, their different proportions, high degree of solution.

A detailed comparison of three approaches has been made in [11].

7   Novel Analytical Approach

The idea behind this approach utilizes the following observation: *Almost all—with very few exceptions—critical points of the extremal function have values in some 4-element set* Q. Indeed, the equiripple property claims that a degree n solution has $2n + 2$ alternation points, those in the interior of E inevitably being critical. Their values belong to the set $Q := \pm\theta, \pm1/\theta$ in the settings (1), (2), (3) or $\pm1 \pm \mu$ in setting (4) or ∂F for the projective setting. This number is roughly equal to the total number $2n - 2$ of critical points of a degree n rational function. The number $g - 1$ of exceptional critical points is counted as

$$g - 1 = \sum_{x:\, R(x)\notin Q} B(R, x) + \sum_{x:\, R(x)\in Q} \left[\frac{1}{2} B(R, x) \right], \tag{5}$$

here the summation is taken over points of the Riemann sphere; $[\cdot]$ is the integer part of a number and $B(R, x)$ is the branching number of the holomorphic map R at the point x. The latter value equals zero in all regular points x including simple poles of $R(x)$, or the multiplicity of the critical point of $R(x)$ otherwise.

Mentioned above exceptional property of extremal rational functions may be rewritten in a form of a generalized Pell-Abel functional equation and eventually gives the desired few-parametric representation of the solution [8] for the normalized $Q = \{\pm 1, \pm 1/k(\tau)\}$

$$R(x) = \text{sn}\left(\int_e^x d\zeta + A(e) \,\Big|\, \tau \right). \tag{6}$$

Here $\text{sn}(\cdot|\tau)$ is the *elliptic sine* of the *modulus* τ related to the value of the deviation (depending on the setting it is μ, θ or $\kappa(F)$); $d\zeta$ is a *holomorphic differential* on the unknown beforehand *hyperelliptic curve*

$$M = M(\mathbf{e}) = \{(w, x) \in \mathbb{C}^2 : \quad w^2 = \prod_{s=1}^{2g+2} (x - e_s)\}, \qquad \mathbf{e} = \{e_s\}_{s=1}^{2g+2}. \tag{7}$$

This curve has branching at the points $e \in \mathbf{e}$ where $R(x)$ takes values from the exceptional set Q *with odd multiplicity*. One can show that the genus g of the curve (7) equals to the above defined number (5) of exceptional critical points plus 1. The arising surface is not arbitrary: it bears a holomorphic differential $d\zeta$ whose periods lie in the rank 2 periods lattice of elliptic sine. The phase shift $A(e)$ is some quarter period of $\text{sn}(\cdot|\cdot)$.

Algebraic curves of this type are not new to mathematicians: they are so-called *Calogero-Moser curves* and describe the dynamics of points on a torus interacting with the Weierstrass potential $\wp(u)$.

8 Examples of Filter Design

We give several examples of optimal magnitude response functions from different classes, all of them are computed by Sergei Lyamaev. Figure 4 shows the solution of fourth Zolotarëv problem with the set E consisting of 30 bands. The solution contains no poles in the transition bands and may be transformed to the transfer

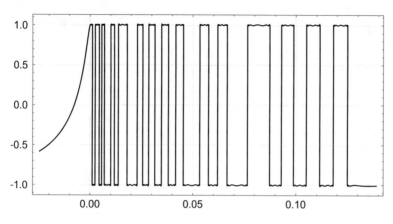

Fig. 4 Optimal magnitude response function with 30 work bands

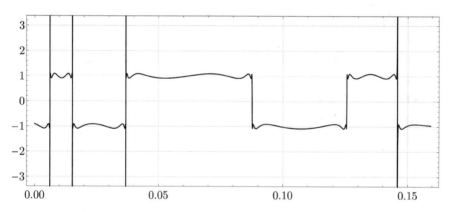

Fig. 5 The minimizer for the fourth Zolotarëv problem on 7 bands with class Σ admitting poles in some transition bands

function of the multiband filter. Figure 5 shows the solution of the problem with seven working bands. Its class Σ admits poles in the transitions and the function cannot be used for the filter synthesis, which does not exclude other possible applications.

Figures 6 and 7 represents a magnitude response function of the so-called double notch filter which eliminates input signal in the narrow vicinities of two given frequencies. Shown here optimal filter has degree $n = 16$ while same specification composite filter has degree $n = 62$.

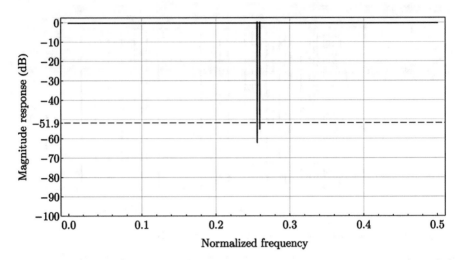

Fig. 6 Optimal *double notch* filter eliminating noise at two given frequencies. Log scale on the vertical axis (amplification)

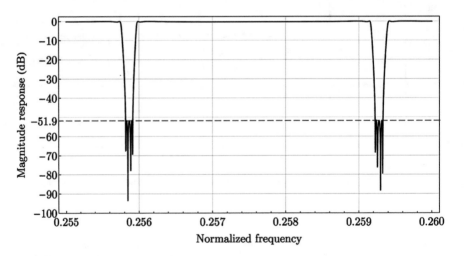

Fig. 7 Magnification of the previous figure at two cutoff frequencies

Acknowledgements Supported by RScF grant 16-11-10349.

References

1. Achïeser, N.I.: Sur un problème de E. Zolotarëv. Bull. Acad. Sci. de l'URSS **VII sér.**(10), 919–931 (1929)
2. Achïeser, N.I.: Elements of the Theory of Elliptic Functions (Russian). Leningrad (1948). Translated from the second Russian edition by H.H. McFaden. Translations of Mathematical Monographs, vol. 79, viii+237 pp. American Mathematical Society, Providence, RI (1990)

3. Amer, R.A.-R., Schwarz, H.R.: Contributions to the approximation problem of electrical filters. Mitt. Inst. Angew. Math. (1964), No. 9, Birkhäuser, Basel. 99 pp. See also R.A.-R. Amer's PHD Thesis, ETH, 1964
4. Bogatyrev, A.B.: Effective computation of Chebyshev polynomials for several intervals. Sb. Math. **190**(11), 1571–1605 (1999)
5. Bogatyrev, A.B.: Effective computation of optimal stability polynomials. Calcolo **41**(4), 247–256 (2004)
6. Bogatyrev, A.B.: Effective solution of the problem of the optimal stability polynomial. Sb. Math. **196**(7), 959–981 (2005)
7. Bogatyrev, A.B.: Ekstremal'nye mnogochleny i rimanovy poverkhnosti. MCCME, Moscow (2005). Translated under the title Extremal Polynomials and Riemann Surfaces, Springer, Berlin (2012)
8. Bogatyrev, A.B.: Chebyshëv representation of rational functions. Mat. Sb. **201**(11), 19–40 (2010) [Sb. Math. (Engl. Transl.) 201(11–12), 1579–1598 (2010)]
9. Bogatyrev, A.B.: Rational functions admitting double decompositions. Trans. Moscow Math. Soc. **73**, 161–165 (2012)
10. Bogatyrev, A.B.: How many Zolotarev fractions are there? Constr. Approx. **46**(1), 37–45 (2017)
11. Bogatyrev, A.B. Goreinov, S.A., Lyamaev, S.Yu.: Analytical approach to multiband filter synthesis and comparison to other approaches. Probl. Inform. Trans. **53**(3), 260–273 (2017). arXiv: 1612.01753
12. Cauer, W.: Theorie der linearen Wechselstromschaltungen, Bd. 1, Leipzig: Becker und Erler, 1941; Bd. 2. Akademie, Berlin (1960)
13. Chu, Q., Wu, X., Chen, F.: Novel compact tri-band bandpass filter with controllable bandwidths. IEEE Microw. Wirel. Compon. Lett. **21**(12), 655–657 (2011)
14. Deslandes, D., Boone, F.: An iterative design procedure for the synthesis of generalized dual-bandpass filters. Int. J. RF Microwave CAE **19**(5), 607–614 (2009)
15. Dubinin, V.N.: An extremal problem for the derivative of a rational function. Math. Notes **100**(5), 714–71 (2016)
16. Dubinin, V.N.: The logarithmic energy of zeros and poles of a rational function. Sib. Math. J. **57**(6), 981–986 (2016)
17. Fuchs, W.J.H.: On Chebyshëv approximation on sets with several components. In: Brannan, D.A., Clunie, J.G. (eds.) Aspects of Contemporary Complex Analysis, pp. 399–408. Academic Press, London, New York (1980)
18. Gonchar, A.A.: The problems of E.I. Zolotarëv which are connected with rational functions. Mat. Sb. **78**(120), no. 4, 640–654 (1969)
19. Gutknecht, M.H., Smith, J.O., Trefethen, L.N.: The Caratheodory-Fejer method for recursive digital filter design. IEEE Trans. Acoust. Speech Signal Process. **31**(6), 1417–1426 (1983)
20. Lee, J., Sarabandi, K.: Design of triple-passband microwave filters using frequency transformation. IEEE Trans. Microw. Theory Tech. **56**(1), 187–193 (2008)
21. Macchiarella, G.: "Equi-ripple" synthesis of multiband prototype filters using a Remez-like algorithm. IEEE Microw. Wirel. Compon. Lett. **23**(5), 231–233 (2013)
22. Malozemov, V.N.: The synthesis problem for a multiband electrical filter. Zh. Vychisl. Mat. i Mat. Fiz. **19**(3), 601–609 (1979)
23. Mohan, A., Singh, S., Biswas, A.: Generalized synthesis and design of symmetrical multiple passband filters. Prog. Electromagn. Res. B **42**, 115–139 (2012)
24. Remez, E.Ya.: Basics of Numerical Methods of Chebyshëv Approximation. Naukova Dumka, Kiev (1969)
25. Veidinger, L.: On the numerical determination of the best approximations in the Chebyshëv sense. Numer. Math. **2**(1), 99–105 (1960)
26. Zolotarëv, E.I.: Application of elliptic functions to questions on functions deviating least and most from zero. Zap. Imp. Akad. Nauk St. Petersburg **30**(5), 1–71 (1877)

Impact of Error in Parameter Estimations on Large Scale Portfolio Optimization

Valery A. Kalyagin and Sergey V. Slashchinin

1 Introduction

Mathematical programming is often concerned with determining the optimum of some real-world objective. One of the main questions in this field is the way of addressing uncertainty. Whereas deterministic optimization models are formulated with known parameters, real life issues almost invariably include uncertain parameters which are unknown at the time when a decision should be made.

An example of this is the area of financial optimization, since there are various uncertainties, such as prices of goods, economic factors, asset returns, turnover constraints, etc. One of the most known problems in the field is portfolio optimization (PO). The objective is to distribute capital between available investment instruments in the financial market. Examples of these instruments are stocks, bonds, options, and bank deposits. The aim is to maximize or keep at the desired level the wealth resulting from the investment, but at the same time minimize the involved risks. The major progress started after publication of Harry Markowitz's seminal work [14]. Since the late 1950s, PO has been an active field in finance. An extensive overview of modern portfolio theory, concepts, and mathematical models for financial markets can be found in the work of Lyuu [13]. The author presents a variety of algorithms for computational techniques in pricing, risk management, and portfolio selection, analyzes their efficiency, and offers thorough theoretical grounding for the proposed models.

One of the ways to deal with uncertainty is to estimate the unknown parameters. Since some optimization models require statistical estimators of parameters, an

V. A. Kalyagin (✉) · S. V. Slashchinin
Laboratory of Algorithms and Technologies for Network Analysis, National Research University
Higher School of Economics, Nizhny Novgorod, Russia
e-mail: vkalyagin@hse.ru

© Springer Nature Switzerland AG 2019
I. C. Demetriou, P. M. Pardalos (eds.), *Approximation and Optimization*,
Springer Optimization and Its Applications 145,
https://doi.org/10.1007/978-3-030-12767-1_9

adequate choice of estimator can substantially improve optimization performance. On the other hand, many estimators contain estimation error or bias that most likely to perturb an optimizer. Consequently, it is not enough to wisely choose a model for the considered problem. Another crucial aspect of success is to determine a proper way to estimate model parameters. Use of improper estimation technique and insufficiency of information may lead to that selected portfolios will have poor out-of-sample performance.

The effect of misspecification and estimation errors on optimal portfolio selection has been a point of interest in the scientific community for many years. For a comprehensive overview of the topic and the modern trends in mean–variance portfolio optimization the reader may refer to the paper of Kolm et al. [9]. The main purpose of this study is to investigate how estimators of uncertain parameters can affect the portfolio optimization results in the case of large scale portfolio.

The remainder of this paper is structured as follows. The first chapter includes some basic principles and theoretical background on mean–variance portfolio optimization and also describes different estimators that might be used within portfolio optimization models, and their statistical properties. Special attention is paid to so-called shrinkage estimators, which are claimed to be one of the most effective in some scenarios. The second chapter presents the experiments conducted on out-of-sample performance of selected optimal portfolios. There will be considered different unusual facts, discovered during the study. At the end some future research directions are discussed.

2 Theoretical Background

2.1 Portfolio Optimization

The basics of portfolio selection were presented in the pioneering work of Markowitz [14]. Usually, the problem of portfolio optimization is formulated as follows. Suppose we have capital W_0 that we need to invest. There are N risky assets on the market with returns $R = (R_1, R_2, \ldots, R_N)$ over some period of time (e.g., week, month, and year). R is a random vector with means $\mathbb{E}(R) = [\mathbb{E}(R_1), \mathbb{E}(R_2), \ldots, \mathbb{E}(R_N)]$ and covariance matrix Σ with elements σ_{ij}. We decide on the amount (share) w_i that should be invested in ith asset, $i = 1, \ldots, N$, such that $\sum_{i=1}^{N} w_i = W_0$. Thus we select an investment portfolio $w = (w_1, w_2, \ldots, w_N)$ with return $R_p = w^\top R$ over period of time. The expected rate of portfolio return is $\overline{R_p} = w^\top \mathbb{E}(R)$. Without loss of generality, we can assume that $W_0 = 1$, then w is the vector of relative amounts, invested in each asset.

2.1.1 Markowitz Model and Its Variations

Every investor is supposed to be rational and risk averse. Since that, from two different portfolios with the same return he chooses the portfolio with lower risk and from the portfolios that have the same level of risk he will prefer the one with higher expected outcome.

The main assumption of the model is that $\mathbb{E}(R)$ and Σ are known (or can be estimated).

Markowitz formulated two criteria for portfolio optimality:

1. Effectiveness of portfolio—expected returns $\overline{R_p} = w^\top \mathbb{E}(R) \to max$
2. Risk—variance of portfolio return $\sigma_p^2 = \mathbb{E}\left((R_p - \overline{R_p})^2\right) = w^\top \Sigma w \to min$

Since both criteria can't be optimized simultaneously, two alternative problems emerge:

$$\underset{w}{\text{minimize}} \quad \frac{1}{2} w^\top \Sigma w$$

$$\text{subject to} \quad \sum_{i=1}^{N} w_i = 1, \tag{1}$$

$$w^\top \mathbb{E}(R) = \tau$$

and

$$\underset{w}{\text{maximize}} \quad w^\top \mathbb{E}(R)$$

$$\text{subject to} \quad \sum_{i=1}^{N} w_i = 1, \tag{2}$$

$$w^\top \Sigma w = v,$$

where v and τ are the predetermined by investor maximum level of risk and the minimum rate of return.

Extra constraint can be used to forbid short selling:

$$w_i \geq 0, \ i = 1, \ldots, N. \tag{3}$$

The Markowitz problem in the form (1) can be efficiently solved using Lagrange multipliers. The weights w_i and the two Lagrange multipliers λ_1 and λ_2 for an optimal portfolio should hold:

$$\sum_{j=1}^{N} \sigma_{ij} w_i - \lambda_1 \mathbb{E}(R_i) - \lambda_2 = 0, \ i = 1, \ldots, N$$

$$\sum_{i=1}^{N} w_i \mathbb{E}(R_i) = \tau \tag{4}$$

$$\sum_{i=1}^{N} w_i = 1.$$

The optimal solution for this system of $N + 2$ equations is

$$w = \frac{C - \tau B}{AC - B^2} \Sigma^{-1} \mathbb{E}(R) + \frac{\tau A - B}{AC - B^2} \Sigma^{-1} \mathbf{1}_N, \tag{5}$$

where $\mathbf{1}_N$ denotes a N-dimensional vector of ones, $A = \mathbb{E}(R^{\mathrm{T}}) \Sigma^{-1} \mathbb{E}(R)$, $B = \mathbf{1}_N^{\mathrm{T}} \Sigma^{-1} \mathbb{E}(R)$, $C = \mathbf{1}_N^{\mathrm{T}} \Sigma^{-1} \mathbf{1}_N$.

In the case of the form (2), when the goal is to achieve the highest return for a given level of risk, sophisticated quadratic programming techniques are needed to find the optimal portfolio.

Since solutions of these optimization problems cannot be dominated (surpassed by two criteria simultaneously by any other portfolio), they are Pareto-efficient and form the **efficient frontier**, which is also called the efficient set. All other portfolios are termed inefficient.

It should be noted that variance is not a very good risk measure, because it penalizes both positive and negative deviations, i.e., profits and losses, since it is a measure of the dispersion of the random variable around the mean. In reality, investors are only interested in minimizing the possibility of insufficient portfolio returns. Consequently, other ways to evaluate risks were proposed (see [2]):

1. Semivariance
   $$SV = \mathbb{E}\left((\{R_p - \overline{R}_p\}^-)^2 \right),$$
   where $\{\bullet\}^-$ denotes negative part or 0, if expression inside the brackets is non-negative.
2. Probability of falling below some level
   $$P_d = Prob(R_p < d).$$
3. The level of return, value-at-risk (VaR), such that $Prob(R_p < VaR)$ is equal to the given level α
   $$VaR_\alpha = \inf\{R_p \in (-\infty, +\infty) : Prob\left(R_p < VaR\right) = \alpha\}.$$
4. Conditional value-at-risk (CVaR) for level α
   Let $R_p^* = \begin{cases} R_p, & if\ R_p \leq VaR_\alpha \\ 0, & if\ VaR < R_p \end{cases}$
   Then $CVaR_\alpha = \mathbb{E}(R_p^*) = \mathbb{E}\left(R_p \mid R_p \leq VaR_\alpha\right).$

Thus, other portfolio optimization problems can be formulated:

1. Chance constrained:

$$\underset{w}{\text{maximize}} \quad w^{\mathrm{T}} \mathbb{E}(R)$$

$$\text{subject to} \quad \sum_{i=1}^{N} w_i = 1,$$

$$Prob\left(R_p < d\right) \leq \alpha.$$

2. With use of VaR or CVaR:

$$\underset{w}{\text{maximize}} \quad w^{\top} \mathbb{E}(R)$$

$$\text{subject to} \quad \sum_{i=1}^{N} w_i = 1,$$

$$\text{CVaR}_{\alpha} \leq d,$$

where α and d are predetermined constants. Solutions of these portfolio optimization problems also form efficient frontiers.

Markowitz theory provides a simple yet powerful framework for controlling the risk–return trade-off of the portfolios. The significant problem for mean–variance model is that too much parameters need to be estimated: N for expected return rates of the assets and $N(N + 1)/2$ for their covariances. Asset returns may be often explained by much less random variables called factors, and different models were developed to exploit this fact.

2.1.2 Single-Factor Model

The first factor model was formulated by Sharpe [17]. It uses only one factor f, which is a random quantity similar to the return on a stock index for the given period. All asset returns depend on this factor in the following way:

$$R_i = \alpha_i + \beta_i f + \varepsilon_i, \tag{6}$$

where α_i and β_i are constants, $\varepsilon = (\varepsilon_1, \varepsilon_2, \ldots,$ and $\varepsilon_N)$—residuals, which are also random. The model assumes that residuals have zero means and are uncorrelated with factor returns and with each other:

$$\text{Cov}(f, \varepsilon_i) = 0, \ i = 1, \ldots, N$$
$$\text{Cov}(\varepsilon_i, \varepsilon_j) = 0, \ i \neq j. \tag{7}$$

There is a total of $3N + 2$ parameters to estimate: α_i, β_i, and σ_i^2, $i = 1, \ldots, N$, $\mathbb{E}(f)$, and σ_f^2, where σ_i^2 denotes the variance of ith residual and σ_f^2 is that of factor return. By σ_{if} we denote the covariance of asset i with market index (factor) returns. The means of asset returns are $\mathbb{E}(R_i) = \alpha_i + \beta_i \mathbb{E}(f)$, and the covariance matrix is $\Phi = \left(\beta\beta^T \sigma_f^2\right) + \Delta$, where Δ is the diagonal covariance matrix of residuals.

2.1.3 Multi-Factor Model

Unfortunately, single-index model is too simple to distinguish all significant market features and trends, while even a small error may lead to negative consequences, even heavy losses of an investor. To partially deal with this problem multi-index models were proposed. It uses $1 < K < N$ random variables (factors) to explain uncertainty of asset returns:

$$R_i = \alpha_i + \sum_{j=1}^{K} \beta_{ij} f_j + \varepsilon_i. \tag{8}$$

The assumptions there are similar to the ones in the single-index model.

The means of asset returns are $\mathbb{E}(R_i) = \alpha_i + \sum_{j=1}^{K} \beta_{ij} \mathbb{E}(f_j)$, and the covariance matrix is $\Phi = (\beta \Sigma_f \beta^T) + \Delta$, where Σ_f is the $K \times K$ covariance matrix of the factors.

One of the main disadvantages of this model is that there is no consensus on the number of factors K that should be used [6]. Also it is hard to understand the nature of these factors, except for the first one, which corresponds to a market index and since that we cannot find theoretical economic interpretation for them.

2.2 Parameters Estimation

2.2.1 Estimation of Means

To solve portfolio optimization problems described in the previous section, model parameters—mean vector and covariance matrix—must be estimated.

Suppose we have some historical data on N stock returns. By X we denote an $N \times T$ matrix of T observations on a system of N random variables $R = R_1, R_2, \ldots, R_N$ representing T returns on N assets, where element X_{ij} is the outcome of ith asset over time period j. We assume that stock returns are independent and identically distributed (IID). This means that each vector $X_j = (X_{1j}, X_{2j}, \ldots, X_{Nj})$, $j = 1..T$, has the same vector of means $\mathbb{E}(R)$ and covariance matrix Σ. Although this assumption almost never holds for the real stocks, most of the estimators for portfolio optimization models use it.

To estimate mean vector $\mathbb{E}(R)$ sample mean is usually used:

$$\overline{X} = \frac{1}{T} X \mathbf{1}_T = \frac{1}{T} \sum_{j=1}^{T} X_j, \tag{9}$$

where $\mathbf{1}_T$ is a $T\mathrm{x}1$ vector of ones. It is unbiased estimator of the true mean:

$$\mathbb{E}\left(\overline{X}\right) = \frac{1}{T}\sum_{j=1}^{T}\mathbb{E}\left(X_j\right) = \frac{1}{T}\left(\mathbb{E}\left(X_1\right) + \mathbb{E}\left(X_2\right) + \cdots + \mathbb{E}\left(X_T\right)\right) =$$

$$= \frac{1}{T}\left(\mathbb{E}\left(R\right) + \mathbb{E}\left(R\right) + \cdots + \mathbb{E}\left(R\right)\right) = \mathbb{E}(R).$$

Also, it is consistent, i.e., converges in probability to the true mean as number T grows

$$\lim_{n\to\infty}\Pr\left(\left|\overline{X} - \mathbb{E}\left(R\right)\right| \geq \epsilon\right) = 0, \text{ for all } \epsilon > 0$$

or

$$\overline{X}\xrightarrow{\text{P}}\mathbb{E}\left(R\right).$$

This follows directly from the law of large numbers in the Khinchin's formulation.

However, despite all the advantages of sample mean, it can be shown that this estimator is inadmissible relative to a quadratic loss function for multivariate normal distribution [18]. Instead of it, several different estimators were proposed, which are based on a statistical technique called **shrinkage**. It was proven that the so-called James–Stein estimator [7] has uniformly lower risk than the sample mean. This estimator is defined as

$$\hat{\mu} = (1 - a)\,\overline{X} + aX_0, \tag{10}$$

where $\hat{\Sigma}$ is some estimator of covariance matrix, X_0—shrinkage target (for example, market average—average of sample means), and a—shrinkage coefficient, which is equal to

$$a_1 = \frac{(N-2)/T}{(\overline{\mu} - \mu_0)^T\hat{\Sigma}^{-1}(\overline{\mu} - \mu_0)}. \tag{11}$$

Using the Bayesian approach, Jorion proposed another form of shrinkage coefficient, which can substantially outperform classical sample mean [8]:

$$a_2 = \frac{(N+2)/T}{N + 2 + (\overline{\mu} - \mu_0)^T\hat{\Sigma}^{-1}(\overline{\mu} - \mu_0)}. \tag{12}$$

2.2.2 Estimation of Covariances

To estimate the true covariance matrix Σ sample covariance matrix can be used:

$$\hat{\Sigma} = \frac{1}{T}\sum_{j=1}^{T}(X_j - \overline{X})(X_j - \overline{X})^T = \frac{1}{T}X\left(\mathbf{I}_{T\mathrm{x}T} - \frac{1}{T}\mathbf{1}_T\mathbf{1}_T^T\right)X^T, \tag{13}$$

where $\mathbf{I}_{TxT}$ is TxT identity matrix. This estimator is biased, but it can be easily corrected by multiplication on term $T/(T-1)$. The main advantages of (13) is that it is maximum likelihood estimator under the assumption of normality of stock returns, and that it is a consistent estimator. But at the same time it has significant drawbacks. One of them is seen from the matrix form: the rank of $\widehat{\Sigma}$ is at most equal to the rank of the matrix $\mathbf{I}_{TxT} - \frac{1}{T}\mathbf{1}_T\mathbf{1}_T^T$, which is $T-1$. This means that the sample covariance matrix is rank-deficient and is not invertible when $N \geq T$. This means, for example, that this estimator cannot be used in (5) to find the exact solution for the portfolio selection problem. Another disadvantage is that estimation error (mean squared error) for the sample covariance matrix is too big and of order N/T; thus, it can considerably perturb mean–variance estimator [10]. For this reasons sample covariance matrix is seldom used in practice.

One of the solutions is to use the single-index model (6) to estimate the true covariance matrix. Suppose that besides historical data on asset returns X we have vector $X_f = (X_{f1}, X_{f2}, \ldots, X_{fT})$ of T observations on some market index (for example, S&P 500). Now we can use it for estimation:

$$\hat{\Phi} = \left(\mathbf{b}\mathbf{b}^T \hat{\sigma}_f^2\right) + \hat{\Delta}, \tag{14}$$

where $\hat{\sigma}_f^2$ is the sample variance of market returns equal to $\frac{1}{T}\sum_{j=1}^T \left(X_{fj} - \overline{X}_f\right)^2$, $\overline{X}_f$ is the sample market mean $\frac{1}{T}\sum_{j=1}^T X_{fj}$, $\hat{\Delta}$ is the sample diagonal covariance matrix of residuals, and $\mathbf{b} = (b_1, b_2, \ldots, b_N)$ is a vector of OLS (ordinary least squares) estimates:

$$b_i = \frac{\sum_{j=1}^T \left(X_{ij} - \overline{X}_i\right)\left(X_{fj} - \overline{X}_f\right)}{\sum_{j=1}^T \left(X_{fj} - \overline{X}_f\right)^2}.$$

Actually, the form of market index is not as important as for CAPM [16]. X_0 should explain a significant part of the variance of stock returns. For this fact, an equal-weighted portfolio can be used to represent market index.

The matrix $\hat{\Phi}$ is non-singular, hence, invertible and can be used in (5). Also, it has lower mean squared error compared to the sample covariance matrix [11].

The main disadvantage is that $\hat{\Phi}$ contains significant bias, which comes from model assumptions (7). For example, if residuals are correlated with each other, then the corresponding covariances will be underestimated or overestimated. Thus, imposing too much structure on the model leads to another significant issue. Multi-factor models can partially solve the problem, since they have less strict structural assumptions. However, as it was noted, it is not clear how much factors should be used.

2.2.3 Ledoit and Wolf Shrinkage Estimator for Covariance Matrix

Another way of finding an optimal trade-off between bias and estimation error was proposed by Ledoit and Wolf [11]. The idea is to take a properly weighted average of the biased estimator (the estimator for singe-index model) and unbiased estimator (the sample covariance matrix), which has greater mean squared error:

$$\alpha \widehat{\Phi} + (1-\alpha) \widehat{\Sigma},$$

where α is a shrinkage intensity constant.

Their approach is supported by the following assumptions:

1. Observations of stock returns are independent and identically distributed (iid) through time. This assumption is similar to the one noted above.
2. The number of assets N is fixed and $N < +\infty$, while T tends to infinity.
3. Asset returns have finite fourth moments.
4. $\Phi \neq \Sigma$.
5. $\sigma_f^2 > 0$.

To derive the optimal shrinkage intensity, they considered the loss function based on the Frobenius norm of the difference between the estimator and the true covariance matrix:

$$L(\alpha) = \left\| \alpha \widehat{\Phi} + (1-\alpha) \widehat{\Sigma} - \Sigma \right\|^2 = \sum_{i=1}^{N} \sum_{j=1}^{N} \left(\alpha \widehat{\phi}_{ij} + (1-\alpha) \widehat{\sigma}_{ij} - \sigma_{ij} \right)^2.$$

By $\widehat{\phi}_{ij}$ we denote i,jth element of the estimator $\widehat{\Phi}$, $\widehat{\sigma}_{ij}$ is i,jth element of $\widehat{\Sigma}$, and σ_{ij} is the true covariance between assets i and j. From this arises risk function:

$$Risk(\alpha) = \mathbb{E}[L(\alpha)] = \sum_{i=1}^{N} \sum_{j=1}^{N} \mathbb{E}\left[\left(\alpha \widehat{\phi}_{ij} + (1-\alpha) \widehat{\sigma}_{ij} - \sigma_{ij} \right)^2 \right]$$

$$= \sum_{i=1}^{N} \sum_{j=1}^{N} \mathbb{E}\left[\left(\alpha \widehat{\phi}_{ij} + (1-\alpha) \widehat{\sigma}_{ij} \right)^2 - 2 \left(\alpha \widehat{\phi}_{ij} + (1-\alpha) \widehat{\sigma}_{ij} \right) \sigma_{ij} + \sigma_{ij}^2 \right]$$

$$= \sum_{i=1}^{N} \sum_{j=1}^{N} \left(\mathbb{E}\left[\left(\alpha \widehat{\phi}_{ij} + (1-\alpha) \widehat{\sigma}_{ij} \right)^2 \right] - \left(\mathbb{E}\left[\alpha \widehat{\phi}_{ij} + (1-\alpha) \widehat{\sigma}_{ij} \right] \right)^2 \right.$$

$$\left. + \left(\mathbb{E}\left[\alpha \widehat{\phi}_{ij} + (1-\alpha) \widehat{\sigma}_{ij} \right] \right)^2 - \mathbb{E}\left[2 \left(\alpha \widehat{\phi}_{ij} + (1-\alpha) \widehat{\sigma}_{ij} \right) \sigma_{ij} \right] + \mathbb{E}\left[\sigma_{ij}^2 \right] \right)$$

$$= \sum_{i=1}^{N} \sum_{j=1}^{N} \left(\mathrm{Var}\left[\alpha \widehat{\phi}_{ij} + (1-\alpha) \widehat{\sigma}_{ij} \right] + \left(\mathbb{E}\left[\alpha \widehat{\phi}_{ij} + (1-\alpha) \widehat{\sigma}_{ij} - \sigma_{ij} \right] \right)^2 \right)$$

$$= \sum_{i=1}^{N} \sum_{j=1}^{N} \left(\alpha^2 \mathrm{Var}\left[\widehat{\phi}_{ij}\right] + (1-\alpha)^2 \mathrm{Var}\left[\widehat{\sigma}_{ij}\right] + 2\alpha\,(1-\alpha)\,\mathrm{Cov}\left(\widehat{\phi}_{ij},\widehat{\sigma}_{ij}\right) \right.$$

$$\left. + \alpha^2 (\phi_{ij} - \sigma_{ij})^2 \right).$$

We want to minimize the risk with respect to α. By calculating the first two derivatives of $Risk\,(\alpha)$ we get:

$$Risk'\,(\alpha) = 2 \sum_{i=1}^{N} \sum_{j=1}^{N} \left(\alpha\,\mathrm{Var}\left[\widehat{\phi}_{ij}\right] - (1-\alpha)\,\mathrm{Var}\left[\widehat{\sigma}_{ij}\right] + (1 - 2\alpha)\,\mathrm{Cov}\left(\widehat{\phi}_{ij},\widehat{\sigma}_{ij}\right) \right.$$

$$\left. + \alpha(\phi_{ij} - \sigma_{ij})^2 \right),$$

$$Risk''\,(\alpha) = 2 \sum_{i=1}^{N} \sum_{j=1}^{N} \left(\mathrm{Var}\left[\widehat{\phi}_{ij} - \widehat{\sigma}_{ij}\right] + (\phi_{ij} - \sigma_{ij})^2 \right).$$

By setting the first derivative to 0 and solving this equation for α^* we obtain the formula for optimal shrinkage intensity:

$$\alpha^* = \frac{\sum_{i=1}^{N} \sum_{j=1}^{N} \left(\mathrm{Var}\left[\widehat{\phi}_{ij}\right] - \mathrm{Cov}\left(\widehat{\phi}_{ij},\widehat{\sigma}_{ij}\right) \right)}{\sum_{i=1}^{N} \sum_{j=1}^{N} \left(\mathrm{Var}\left[\widehat{\phi}_{ij} - \widehat{\sigma}_{ij}\right] + (\phi_{ij} - \sigma_{ij})^2 \right)}.$$

Note that because $Risk''\,(\alpha)$ is positive everywhere, this intensity minimizes the risk function.

Let $\widehat{\theta}$ denote an estimator for a parameter vector θ, and $\sqrt{T}(\widehat{\theta} - \theta)$ converges to some distribution with zero mean and covariance matrix V as T tends to infinity. Then $\widehat{\theta}$ has asymptotic distribution with parameters θ, $\frac{1}{T}V$. The term $\frac{1}{T}V$ denotes asymptotic covariance matrix of $\widehat{\theta}$.

The following theorem shows the asymptotically optimal choice of shrinkage constant.

Theorem 1 (Ledoit and Wolf [11])

$$\alpha^* = \frac{1}{T}\frac{\pi - \rho}{\gamma} + O\left(\frac{1}{T^2}\right)$$

By π we denote the sum of asymptotic variances of the elements of the matrix $\widehat{\Sigma}$, multiplied by the term $\sqrt{T}$: $\pi = \sum_{i=1}^{N} \sum_{j=1}^{N} \mathrm{AsyVar}\left(\widehat{\sigma}_{ij}\right)$. By ρ we denote the sum of asymptotic covariances of the elements of the matrix $\widehat{\Sigma}$ and with the elements of the matrix $\widehat{\Phi}$ multiplied by the term $\sqrt{T}$: $\rho = \sum_{i=1}^{N} \sum_{j=1}^{N} \mathrm{AsyCov}\left(\widehat{\phi}_{ij},\widehat{\sigma}_{ij}\right)$.

Finally γ is the measure of the misspecification of the single-index model: $\gamma = \sum_{i=1}^{N} \sum_{j=1}^{N} \left(\phi_{ij} - \sigma_{ij} \right)^2$. *Thus, asymptotically optimal shrinkage intensity is*

$$\alpha^* = \frac{k}{T},$$

where $k = \frac{\pi - \rho}{\gamma}$. The next theorem helps to find a consistent estimator for k

Theorem 2 (Ledoit and Wolf [11]) $\hat{k} = \frac{p-r}{c}$ *is a consistent estimator for k. By p we denote* $\sum_{i=1}^{N} \sum_{j=1}^{N} p_{ij}$, *where a consistent estimator of* $\mathrm{AsyVar}\left(\widehat{\sigma}_{ij}\right)$ *is denoted by* $p_{ij} = \frac{1}{T} \sum_{t=1}^{T} \left\{ \left(X_{it} - \overline{X}_i \right) \left(X_{jt} - \overline{X}_j \right) - \widehat{\sigma}_{ij} \right\}^2$. *By r we denote* $\sum_{i=1}^{N} \sum_{j=1}^{N} r_{ij}$, *where r_{ij} is a consistent estimator for* $\mathrm{AsyCov}\left(\widehat{\phi}_{ij}, \widehat{\sigma}_{ij}\right)$ *and $r_{ij} = \frac{1}{T} \sum_{t=1}^{T} r_{ijt}$ with*

$$r_{ijt} = \frac{\widehat{\sigma}_{jf}\sqrt{\widehat{\sigma}_f^2}\left(X_{it} - \overline{X}_i \right) + \widehat{\sigma}_{if}\sqrt{\widehat{\sigma}_f^2}\left(X_{jt} - \overline{X}_j \right) + \widehat{\sigma}_{if}\widehat{\sigma}_{jf}\left(X_{ft} - \overline{X}_f \right)}{\widehat{\sigma}_f^2}$$

$$\left(X_{ft} - \overline{X}_f \right)\left(X_{it} - \overline{X}_i \right)\left(X_{jt} - \overline{X}_j \right) - \widehat{\phi}_{ij}\widehat{\sigma}_{ij}.$$

Finally, $c = \sum_{i=1}^{N} \sum_{j=1}^{N} c_{ij}$, where c_{ij} is a consistent estimator of $\left(\phi_{ij} - \sigma_{ij}\right)^2$ equal to $\left(\widehat{\phi}_{ij} - \widehat{\sigma}_{ij}\right)^2$.

The corresponding optimal estimator for the true covariance matrix is

$$S = \frac{\hat{k}}{T}\widehat{\Phi} + \left(1 - \frac{\hat{k}}{T} \right)\widehat{\Sigma}. \tag{15}$$

Detailed proofs for Theorems 1 and 2 are given in Ledoit and Wolf [11]. The authors claim that "nobody should be using the sample covariance matrix for the purpose of portfolio optimization" [12]. They show empirical evidence of their estimator performance. This estimator became industry-standard benchmark for estimating the covariance matrix of stock returns and was implemented inside many software packages for Python, Matlab, and R.

This and other estimators will be used in the next chapter for numerical experiments.

3 Properties of Selected Portfolios

This part consists of empirical analysis of selected portfolios resulting from solving optimization problems and how estimators (which were described above) can affect their out-of-sample performance. To show this more explicitly, Markowitz's minimal risk model (1) has been chosen, because its effectiveness directly depends on how good uncertain parameters (vector of means and covariance matrix of stock returns) are estimated. So, we solve portfolio optimization problem in the form of (1) both with and without constraint of short sales (3). Desired level of annual portfolio return τ chosen to be at 0.2.

To solve the problem, we use the following:

1. When we have no constraints on short sales, we find exact solution in the form of (5).
2. With constraints on short sales we use sequential quadratic programming algorithm [15] and interior point algorithm for quadratic programming [1], implemented in Matlab Optimization Toolbox package.

The framework for experiments and optimization problem solver were implemented using Matlab 2016a. All computations were done on a machine with Intel Core i7 2.20 GHz processor and 8 Gb RAM running Windows 10 operating system.

In this section we will use sample mean to estimate $\mathbb{E}(R)$ and the following estimators for Σ during the experiments:

1. Shrinkage to single-index matrix (15), denoted as "Ledoit1."
2. Shrinkage to constant correlation matrix, denoted as "Ledoit2."
3. The sample covariance matrix (13), denoted as "Sample."
4. The single-index model estimator (14), denoted as "Single ind."
5. The estimator for the multi-index model estimator, denoted as "Multi."
6. The sample covariance matrix with constant correlations, denoted as "Const corr."

3.1 Risk of Selected Portfolios

3.1.1 Real Data

The first experiment is aimed to measure effectiveness of selected optimal portfolios in conditions of real market. The methodology is very similar to the one used in the work of Ledoit and Wolf [11]. We collected data on monthly stock returns from January 1984 to December 2016 of 228 assets from NYSE (New York Stock Exchange) using Yahoo Finance service. Then the following procedure was repeated for every year t = 1994, 1995...2016. Using historical data on the previous ten years (t-10, t-9...t-1, 120 observations) we estimate parameters for optimization problem, find optimal portfolios, and measure their performance (in terms of annual return and standard deviation) during 12 months of year t. Thus, the experiment is repeated 23 times. The results are presented in Table 1 for optimization with no constraints on short selling and in Table 2 for the situation, when short sales are forbidden.

Annual return is calculated as the sum of monthly returns during the testing period of 12 months (when the performance of portfolios is measured).

Table 1 Real data from NYSE, short sales are allowed

Year	Estimator					
	Ledoit1	Ledoit2	Sample	Single ind	Multi	Const corr
(a) Annualized standard deviation						
1994	0.1181	0.1240	0.1123	0.1360	0.1228	0.1505
1995	0.0651	0.0832	0.1167	0.1075	0.0589	0.1315
1996	0.1034	0.0970	0.1212	0.1384	0.1141	0.1449
1997	0.1163	0.1343	0.1122	0.1465	0.1402	0.1721
1998	0.1253	0.1127	0.1938	0.1189	0.1099	0.1196
1999	0.1355	0.1584	0.1634	0.1575	0.1259	0.1793
2000	0.0937	0.1070	0.1469	0.1665	0.1021	0.2137
2001	0.1056	0.1099	0.1818	0.1405	0.1164	0.1686
2002	0.1457	0.1367	0.2181	0.1369	0.1464	0.1447
2003	0.0525	0.0744	0.0851	0.0724	0.0546	0.0990
2004	0.0862	0.0937	0.1422	0.0938	0.0856	0.1016
2005	0.0967	0.1061	0.1256	0.0790	0.0767	0.1001
2006	0.0685	0.0582	0.1073	0.0666	0.0689	0.0545
2007	0.1034	0.1075	0.1336	0.1083	0.1110	0.1065
2008	0.1320	0.1235	0.2061	0.1225	0.1443	0.1214
2009	0.1739	0.1861	0.2347	0.2117	0.1763	0.2354
2010	0.0994	0.1028	0.1258	0.0836	0.0819	0.0917
2011	0.0758	0.0825	0.0969	0.0720	0.0760	0.0896
2012	0.0821	0.0985	0.1242	0.0972	0.0874	0.1379
2013	0.1094	0.1190	0.1055	0.1378	0.1334	0.1698
2014	0.0781	0.0817	0.0851	0.0927	0.0827	0.1084
2015	0.1141	0.1192	0.1288	0.1328	0.1237	0.1677
2016	0.1296	0.1333	0.0985	0.1530	0.1425	0.1947
Average	0.1048	0.1109	0.1377	0.1205	0.1079	0.1393
(b) Mean annual return						
1994	−0.0461	−0.0795	0.0560	−0.0656	−0.0375	−0.0986
1995	0.2375	0.2529	0.2802	0.2974	0.2218	0.3226
1996	0.1380	0.1386	0.1538	0.0538	0.0831	0.0200
1997	0.1851	0.1870	0.1854	0.2191	0.1668	0.2431
1998	0.2204	0.2674	0.2350	0.2062	0.1946	0.3037
1999	−0.1526	−0.1768	−0.1104	−0.1508	−0.1030	−0.1519
2000	0.1002	0.0638	0.1460	0.2963	0.1410	0.2725
2001	−0.0245	0.0143	−0.0129	−0.0720	−0.0567	−0.0640
2002	−0.0911	−0.0874	−0.1861	−0.1276	−0.1187	−0.1240
2003	0.1591	0.1204	0.2573	0.0921	0.1352	0.0270
2004	0.0954	0.0967	0.1537	0.0852	0.0954	0.0704
2005	0.0344	0.0040	0.1377	0.0430	0.0392	−0.0249
2006	0.1802	0.2069	0.2414	0.1766	0.1656	0.1826
2007	0.0307	0.0103	0.0195	0.0434	−0.0209	0.0397

(continued)

Table 1 (continued)

Year	Estimator					
	Ledoit1	Ledoit2	Sample	Single ind	Multi	Const corr
2008	−0.0501	−0.0030	−0.2762	0.0204	−0.0749	0.0509
2009	−0.0786	−0.0679	−0.0519	−0.1569	−0.1211	−0.1910
2010	0.1400	0.1163	0.1717	0.1255	0.1498	0.0922
2011	0.2080	0.2523	0.1083	0.2242	0.2160	0.3006
2012	0.0931	0.0684	0.0558	0.0363	0.0586	−0.0079
2013	0.1689	0.1902	0.1749	0.1418	0.1456	0.1400
2014	0.1855	0.2171	0.1960	0.1996	0.2239	0.2605
2015	0.1829	0.2294	0.2718	0.1109	0.1253	0.1598
2016	0.0289	0.0220	0.1227	0.0380	0.0436	0.0270
Average	0.0846	0.0888	0.1013	0.0799	0.0727	0.0804

Deviation $\mathbb{D}$ is measured as

$$\mathbb{D} = \sqrt{\sum_{i=1}^{12} \left(r_{pi} - \frac{\tau}{12}\right)^2},$$

where r_{pi} denotes return of the selected portfolio over testing month i.

As it seen from Table 1a, portfolios selected with use of shrinkage estimators have considerably lower deviation compared to the others. So, in this case, experiment shows results which are similar to the results in the work of Ledoit and Wolf [11]. At the same time, Table 1b demonstrates that when we have constraints on short sales this advantage of shrinkage estimators of covariance matrix almost disappears.

It also worth noting that both with and without constraints annual returns of selected portfolios do not reach the desired level of 0.2 on average. It might be due to nonstationarity of the market, while our model assumes that parameters do not change over time.

3.1.2 Generated Data

This time to evaluate how good selected portfolios perform out-of-sample we generate data from multivariate normal distribution with mean vector and covariance matrix equal to the sample means and the sample covariance matrix estimated over whole period of 33 years from 1984 to 2016. The number of observations in generated sample for parameter estimation is still 120 and to test effectiveness of portfolios we generate 12 observations. For more accurate results, which are presented in Table 3, experiment is repeated 500 times, then average over repetitions is taken. These results demonstrate the same features that were seen in the previous experiment—the significant advantage of shrinkage estimators in terms of standard

Table 2 Real data from NYSE, short sales are forbidden

Year	Estimator					
	Ledoit1	Ledoit2	Sample	Single ind	Multi	Const corr
(a) Annualized standard deviation						
1994	0.1247	0.1273	0.1208	0.1413	0.1156	0.1318
1995	0.0652	0.0706	0.0717	0.0703	0.0590	0.0925
1996	0.1078	0.1034	0.0987	0.1187	0.1119	0.1126
1997	0.1472	0.1613	0.1403	0.1524	0.1408	0.1750
1998	0.1617	0.1588	0.1753	0.1533	0.1651	0.1394
1999	0.1335	0.1237	0.1181	0.1551	0.1364	0.1252
2000	0.1327	0.1299	0.1376	0.1405	0.1294	0.1330
2001	0.1127	0.1049	0.1124	0.1250	0.1235	0.1350
2002	0.1522	0.1507	0.1730	0.1523	0.1455	0.1684
2003	0.1044	0.0960	0.1115	0.0991	0.1059	0.1000
2004	0.0952	0.0917	0.0967	0.0957	0.0944	0.0815
2005	0.0893	0.0944	0.0854	0.0948	0.0922	0.0911
2006	0.0745	0.0702	0.0718	0.0831	0.0845	0.0816
2007	0.1427	0.1393	0.1459	0.1478	0.1441	0.1358
2008	0.2688	0.2903	0.2712	0.2727	0.2727	0.3233
2009	0.1994	0.2002	0.2190	0.1877	0.2041	0.1900
2010	0.1486	0.1600	0.1531	0.1469	0.1484	0.1667
2011	0.1683	0.1750	0.1733	0.1677	0.1670	0.1780
2012	0.1276	0.1228	0.1200	0.1343	0.1246	0.1276
2013	0.1362	0.1381	0.1317	0.1394	0.1355	0.1417
2014	0.1026	0.1064	0.1025	0.1036	0.0973	0.1112
2015	0.1300	0.1342	0.1346	0.1281	0.1289	0.1326
2016	0.1528	0.1570	0.1565	0.1549	0.1518	0.1633
Average	0.1338	0.1351	0.1357	0.1376	0.1339	0.1408
(b) Mean annual return						
1994	−0.0354	−0.0214	−0.0291	−0.0683	−0.0144	−0.0291
1995	0.2914	0.3306	0.2544	0.3112	0.2634	0.3441
1996	0.1576	0.1345	0.1781	0.1246	0.1730	0.1181
1997	0.2570	0.2950	0.2374	0.2689	0.2353	0.3054
1998	0.1819	0.2396	0.1730	0.1901	0.1740	0.2278
1999	−0.1312	−0.0741	−0.0855	−0.1883	−0.1302	−0.0329
2000	0.1675	0.1109	0.1265	0.2047	0.1463	0.0569
2001	−0.0443	−0.0043	−0.0537	−0.0565	−0.0596	−0.0713
2002	−0.0938	−0.1076	−0.1255	−0.1038	−0.0994	−0.1411
2003	0.2995	0.2673	0.2990	0.2901	0.3149	0.2290
2004	0.1281	0.1273	0.1177	0.1245	0.1161	0.1492
2005	0.0683	0.0672	0.0668	0.0693	0.0776	0.0601
2006	0.1623	0.1783	0.1514	0.1618	0.1716	0.2097
2007	−0.0502	−0.0207	−0.0667	−0.0550	−0.0478	0.0216

(continued)

Table 2 (continued)

| Year | Estimator | | | | | |
	Ledoit1	Ledoit2	Sample	Single ind	Multi	Const corr
2008	−0.3462	−0.3607	−0.3440	−0.3459	−0.3647	−0.4043
2009	0.2001	0.2132	0.1916	0.2139	0.1818	0.2240
2010	0.2506	0.2084	0.2856	0.2363	0.2391	0.1924
2011	0.0632	0.0933	0.0766	0.0536	0.0769	0.0778
2012	0.1770	0.1314	0.1975	0.1726	0.1691	0.0908
2013	0.1795	0.1691	0.1657	0.1803	0.1831	0.1502
2014	0.1035	0.1101	0.0980	0.1024	0.1253	0.0988
2015	0.1442	0.1187	0.1495	0.1398	0.1515	0.1080
2016	−0.1054	−0.1085	−0.1073	−0.1085	−0.0996	−0.1162
Average	0.0880	0.0912	0.0851	0.0834	0.0862	0.0813

Table 3 Generated samples

Estimator	Ledoit1	Ledoit2	Sample	Single ind	Multi	Const corr	True
(a) Short sales are allowed							
STD Deviation	0.0873	0.0925	0.1088	0.1154	0.0978	0.1410	0.0556
Annual mean return	0.1304	0.1324	0.1437	0.1341	0.1382	0.1350	0.2004
(b) Short sales are forbidden							
STD Deviation	0.1124	0.1160	0.1163	0.1185	0.1159	0.1481	0.1911
Annual mean return	0.1139	0.1170	0.1141	0.1134	0.1193	0.1148	0.2003

deviation, which vanishes when we impose short selling constraints. But a much more curious and unexpected fact is that even with generated data, when true parameters do not change, portfolios still do not reach the needed level of return on average. Returns of selected portfolios have significant bias towards zero related to the true optimal portfolio. So, this phenomenon doesn't disappear on generated data and cannot be explained by nonstationarity of the market. Thus, it demands further research to understand its nature.

3.2 Bias of Portfolio Returns

Asymptotic Behavior The first experiment in this section is intended to study how the phenomenon of bias depends on number of observations in the sample that we use for estimation and how it changes as this number grows. Now we consistently increase the size T of sample from 120 to 540 (we still test performance of portfolios on 12 observations) with a step of 60 observations. The results are shown in Table 4.

Table 4 Dependence on sample size T

T	Estimator		
	Ledoit1	Sample	Single ind
(a) Annualized standard deviation, short sales are allowed			
120	0.0918	0.1132	0.1255
180	0.0873	0.1432	0.1292
240	0.0812	0.2289	0.1301
300	0.0793	0.1103	0.1311
360	0.0758	0.0891	0.1356
420	0.0740	0.0806	0.1344
480	0.0733	0.0769	0.1357
540	0.0711	0.0725	0.1369
(b) Annualized standard deviation, short sales are forbidden			
120	0.1194	0.1216	0.1248
180	0.1246	0.1256	0.1308
240	0.1271	0.1277	0.1315
300	0.1356	0.1360	0.1399
360	0.1382	0.1384	0.1420
420	0.1403	0.1405	0.1440
480	0.1460	0.1461	0.1496
540	0.1517	0.1518	0.1542
(c) Mean annual return, short sales are allowed			
120	0.1333	0.1388	0.1406
180	0.1387	0.1390	0.1483
240	0.1426	0.1443	0.1441
300	0.1441	0.1586	0.1591
360	0.1517	0.1653	0.1531
420	0.1568	0.1654	0.1660
480	0.1560	0.1696	0.1663
540	0.1602	0.1723	0.1629
(d) Mean annual return, short sales are forbidden			
120	0.1283	0.1260	0.1321
180	0.1384	0.1379	0.1378
240	0.1281	0.1285	0.1271
300	0.1286	0.1282	0.1283
360	0.1465	0.1468	0.1452
420	0.1451	0.1444	0.1479
480	0.1558	0.1560	0.1557
540	0.1610	0.1606	0.1628

It is evident that the larger the sample size grows, the less noticeable the bias of portfolio returns becomes. Also, if we generate a large sample of size 2000 and more, the phenomenon will almost disappear (average returns will be at level of 0.198 in case of use of the sample covariance matrix). The better we estimate parameters, the closer we get to the desired level of return. This fact leaves us with another question—Is it error in estimation of means or covariance matrix that affects the bias the most?

To answer this question, two similar procedures were carried out (Tables 5a,b,c,d). In the first one, the number of observations to estimate covariance matrix was 120, while for estimation of means the sample size differed from 120 to 540, plus one stage, where the real market data (on returns over 396 months) were used. In the other, the size of sample for estimation of means was constantly 120, while for covariance matrix this number has consistently changed.

The results demonstrate two important facts:

1. The bias depends mostly on accuracy of estimation of the true mean vector (accuracy of covariance matrix estimation almost doesn't affect bias in case of constraints on short sales). This result reproduces the findings of Best and Grauer [4] and Chopra and Ziemba [5], which show that effect from the estimation error in the expected returns usually has more influence than in covariance matrix for mean–variance optimization.
2. The increase from 120 in sample size for covariance matrix entails considerable decrease of portfolios standard deviation when short selling is allowed and has almost insignificant effect in situation with the constraint.

Consequently, for better understanding of the phenomenon it is essential to study how it depends on vector of means.

Dependence on the Mean Vector Figure 1 shows the histogram of mean returns of all assets. It is seen that only few assets have mean returns around desired level of 0.2. The first experiment in this section is focused on how bias changes with change of desired level τ from 0.12 to 0.2 (Tables 6a,b).

So, the bias becomes noticeable for $\tau = 0.14$ and above. We get the similar results when we conduct analogical experiment. This time, instead of changing desired level of returns, we just increase all elements of mean vector of annual returns by 0.03. The results are presented in Table 6c.

The true nature of bias can be seen in Figure 2a and b. These graphs show how the average annual returns of selected portfolios change with change of target level τ the US market (Figure 2a) and German market (Figure 2b, Frankfurt Stock Exchange, number of assets N=150, all assets that were presented on the market during the 180 months period of 2002–2016) for different estimators of covariances (sample covariance matrix, single-index estimator, Ledoit and Wolf estimator, constant correlation matrix, and performance of the true optimal portfolio). It should be noted that the experiments show similar results for the different markets (Euronext Paris Stock Exchange and London Stock Exchange), which indicates that the phenomenon of bias is not due to features of a certain stock market.

Table 5 Dependence on estimation accuracy for different parameters

T	Estimator		
	Ledoit1	Sample	Single ind
(a) Dependence on sample size T for means, mean annual return, short sales are allowed			
120	0.1321	0.1304	0.1332
180	0.1338	0.1334	0.1371
240	0.1413	0.1510	0.1489
300	0.1542	0.1513	0.1501
360	0.1513	0.1587	0.1593
420	0.1534	0.1628	0.1636
480	0.1577	0.1666	0.1573
540	0.1650	0.1689	0.1711
Real market data	0.1926	0.1953	0.1842
(b) Dependence on sample size T for means, mean annual return, short sales are forbidden			
120	0.1271	0.1269	0.1292
180	0.1285	0.1254	0.1302
240	0.1291	0.1297	0.1284
300	0.1313	0.1311	0.1323
360	0.1408	0.1403	0.1402
420	0.1445	0.1464	0.1427
480	0.1504	0.1499	0.1504
540	0.1529	0.1553	0.1520
Real market data	0.1929	0.1933	0.1924
(c) Dependence on sample size T for covariance matrix, mean annual return, short sales are allowed			
120	0.1282	0.1338	0.1370
180	0.1329	0.1428	0.1360
240	0.1321	0.1483	0.1408
300	0.1451	0.1496	0.1492
360	0.1373	0.1497	0.1436
420	0.1473	0.1494	0.1496
480	0.1390	0.1544	0.1501
540	0.1472	0.1494	0.1489
Real market data	0.1496	0.1554	0.1472
(d) Dependence on sample size T for covariance matrix, mean annual return, short sales are forbidden			
120	0.1198	0.1185	0.1196
180	0.1300	0.1295	0.1275
240	0.1283	0.1278	0.1252
300	0.1286	0.1270	0.1359
360	0.1276	0.1269	0.1267
420	0.1317	0.1308	0.1369
480	0.1289	0.1289	0.1283
540	0.1229	0.1236	0.1220
Real market data	0.1345	0.1339	0.1382

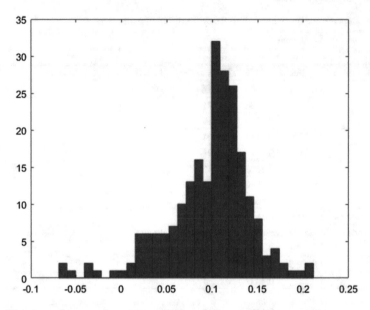

Fig. 1 Histogram of means of asset returns (US market)

The effect of bias is determined by the distribution of expected returns. The further the target level τ from mean (0.12 for the US market, see Figure 1), the more the out-of-sample performance (average annual return) of selected portfolios deviates from what we expect. The figures demonstrate that average returns of selected portfolios are located inside some narrow interval, which is far smaller than [min τ, max τ]. Moreover, it is evident that the form of the chosen estimator only affects where the bounds of this interval are located, not its size. But at the same time, the estimator of covariance matrix controls standard deviation of portfolio returns, thus, its robustness. Figure 2c illustrates how standard deviation of selected portfolio returns from τ depends on target desired level of returns. Ledoit and Wolf estimator has significant advantage compared to the others in terms of robustness, but it is still far from the results for true optimal portfolios due to presence of estimation error for the mean vector.

When there is a lack of assets with desired mean returns, the negative role of estimation error becomes more dramatic and the phenomenon of bias appears. To support this statement, 2 experiments with different distribution of the true mean vector were conducted. In the first one the true mean vector in the generators was replaced by the vector from "2-peaks" distribution (see Figure 3a), which is formed by 2 Normal distributions with means -0.2 and 0.2, and std deviation equal to half of that for the original mean vector. In the second experiment the true means of assets were distributed uniformly from minimum to maximum of the elements of the original vector.

Table 6 Dependence on desired level τ of mean annual return

Tau	Estimator		
	Ledoit1	Sample	Single ind
(a) Short sales are allowed			
0.12	0.1215	0.1282	0.1329
0.13	0.1234	0.1232	0.1353
0.14	0.1274	0.1315	0.1352
0.15	0.1218	0.1225	0.1462
0.16	0.1316	0.1269	0.1487
0.17	0.1334	0.1380	0.1374
0.18	0.1368	0.1363	0.1395
0.19	0.1311	0.1392	0.1328
0.20	0.1345	0.1444	0.1417
(b) Short sales are forbidden			
0.12	0.1036	0.1034	0.1071
0,13	0.1175	0.1224	0.1167
0,14	0.1208	0.1196	0.1178
0.15	0.1215	0.1151	0.1310
0.16	0.1229	0.1227	0.1231
0.17	0.1244	0.1249	0.1240
0.18	0.1188	0.1173	0.1249
0.19	0.1193	0.1141	0.1223
0.20	0.1340	0.1325	0.1347
(c) Average portfolio returns with the "true" mean vector with elements increased by 0.03			
Ledoit1	Sample	Single index	True
Short sales are allowed			
0.1702	0.1687	0.1792	0.2080
Short sales are forbidden			
0.1662	0.1645	0.1724	0.2005

As it is seen from the graphs, the effect of bias disappears when there are enough assets with desired level of returns.

From all the results presented in Figures 2, 3, 4, 5 we can conclude that the effect of bias may appear in situations, when there are only few assets, that reach needed level of annual return, and when the size of sample is considerably small, so estimator of mean return has large variance. When we have a sufficient number of assets with desired level of return or when the sample size is big enough, then the phenomenon fades away.

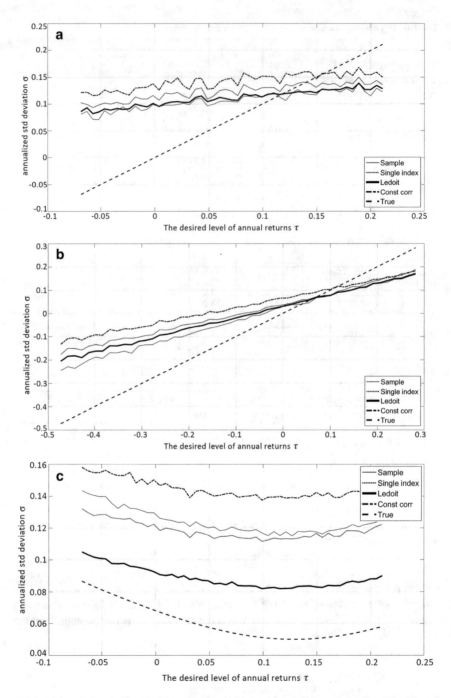

Fig. 2 Dependence of annual returns of selected optimal portfolios on the desired level τ. (**a**) US market, N=228. (**b**) German market, N=150. Dependence of deviation $\mathbb{D}$ of selected optimal portfolios on the desired level τ: (**c**) US market, N=228

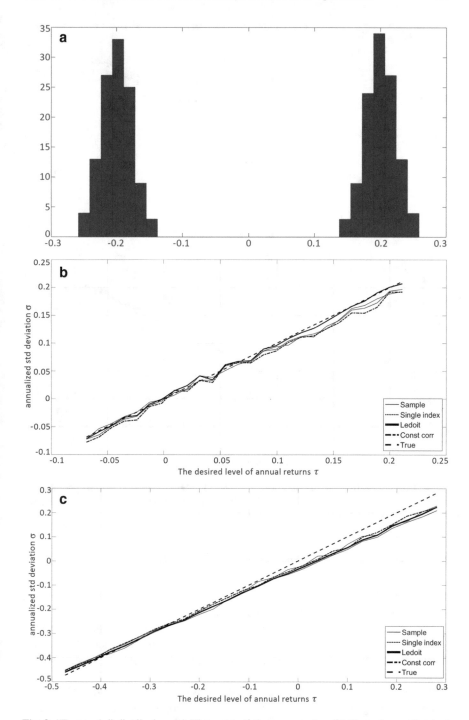

Fig. 3 "Two-peaks" distribution. (**a**) Histogram of the mean vector. (**b**) Dependence of annual returns of selected optimal portfolios on the desired level τ (US market, N=228). (**c**) Dependence of annual returns of selected optimal portfolios on the desired level τ (German market, N=150)

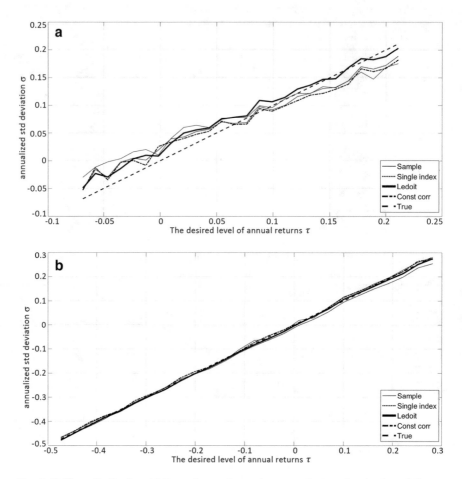

Fig. 4 Uniform distribution. (**a**) Dependence of annual returns of selected optimal portfolios on the desired level τ (US market, N=228). (**b**) Dependence of annual returns of selected optimal portfolios on the desired level τ (German market, N=150)

3.3 Shrinkage Estimators for Mean Vectors

In an attempt to deal with the problem of bias, we tried to use shrinkage estimators for mean vectors [8] instead of traditional sample means.

We measure the performance of the estimators in the same framework with generated samples from multivariate Normal distribution. To generate samples, distribution parameters (sample means and covariance matrix) were estimated on the data (180 monthly observations, i.e., 15 years) from French and German markets (Euronext Paris Stock Exchange and Frankfurt Stock Exchange). To estimate the parameters, we generate a sample set of size $T = 120$, then solve the optimization problem and evaluate the performance of selected portfolios on a new generated test

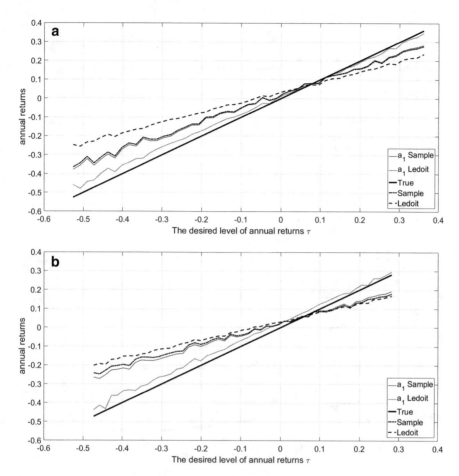

Fig. 5 Dependence of annual returns of selected optimal portfolios on the desired level τ. (**a**) French market, N=108. (**b**) German market, N=150

set with 12 elements (control year). This procedure is repeated for every desired level τ, which changes from the minimum element of the market's true mean vector to the maximum element.

The following notation will be used to specify combinations of estimators for $\mathbb{E}(R)$ and Σ in figures:

1. a_1 **Sample** denotes the combination of shrinkage estimator for means with coefficient a_1 (11) and sample covariance matrix (13).
2. a_2 **Sample** is similar to the previous, but uses shrinkage coefficient a_2 (12).
3. a_1 **Ledoit** denotes the combination of shrinkage estimator for means with coefficient a_1 (11) and Wolf–Ledoit covariance matrix (15), which is also used for estimation of a_1.
4. a_2 **Ledoit** is similar to the previous, but uses shrinkage coefficient a_2 (12).

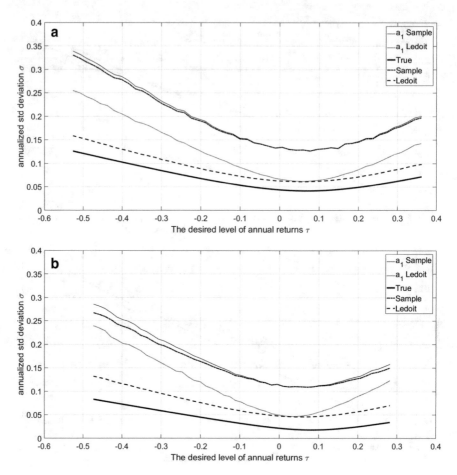

Fig. 6 Dependence of standard deviation σ of selected optimal portfolios on the desired level τ. (**a**) French market, N=108. (**b**) German market, N=150

5. Traditional sample estimators denoted as **Sample**.
6. The combination of sample means (9) and Wolf–Ledoit covariance matrix denoted as **Ledoit**.
7. Label **True** is used for the true parameters, the performance of true optimal portfolios.

3.3.1 Improvements from Shrinkage Estimators for Means (a_1)

The first experiment in this section is aimed to compare the performance of selected portfolios at different desired levels τ of annual returns. The results are presented in Figures 5 and 6.

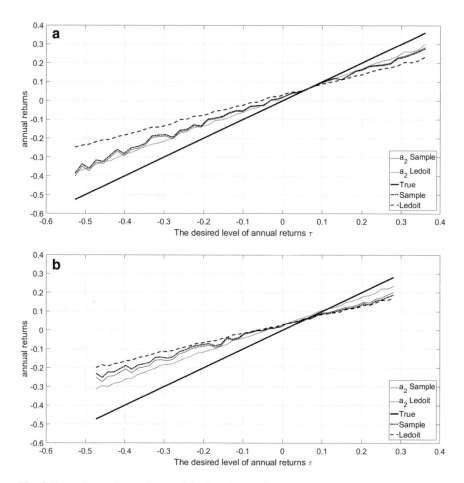

Fig. 7 Dependence of annual returns of selected optimal portfolios on the desired level τ (a_2). (**a**) French market, N=108. (**b**) German market, N=150

The results show considerable improvements in terms of average annual returns of selected portfolios (see Figure 5). For both markets a_1 **equal Ledoit** shows annual returns which are very close to those of the true optimal portfolios, so the phenomenon of bias fades away. However, without Wolf–Ledoit estimator for covariance matrix, the improvement might be not so significant (see Figure 5a).

3.3.2 Improvements from Shrinkage Estimators for Means(a_2)

The cost of this improvement is evident from Figures 7 and 8. With the average returns shrinkage estimators for mean tend to increase the average level of risk of selected portfolios (Figure 7). Although it helps to achieve the desired level τ, the

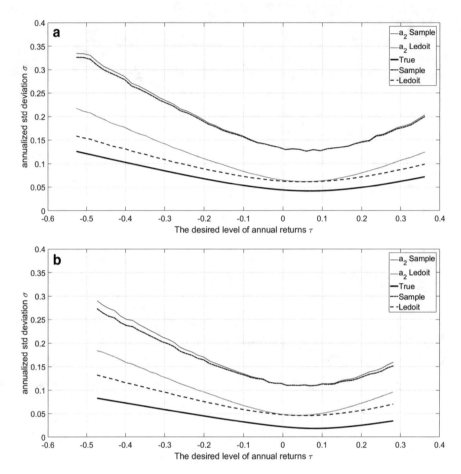

Fig. 8 Dependence of standard deviation σ of selected optimal portfolios on the desired level τ (a_2). (**a**) French market, N=108. (**b**) German market, N=150

selected portfolios can still be far from the true optimal portfolios in terms of risks. But the advantages of Wolf–Ledoit are still present, so the level of risks with this estimator is significantly lower than with sample covariance matrix (Figure 8).

To investigate the reasons of this problem, one might consider the weights of portfolios (Figure 9). The higher norm of vector of weights may lead to higher fluctuations of $w^{\top}\mathbb{E}(R)$, and consequently higher levels of risk. The figure demonstrates that portfolios selected with Wolf–Ledoit estimator usually have lower norm of weights, but may not reach the desired level of annual returns because of it. Also, if we consider the form of the top lines (**a_1 equal Sample**) in Figures 6b and 9b, it is obvious from where this dramatic increase of risk (deviation $\mathbb{D}$) levels did come from.

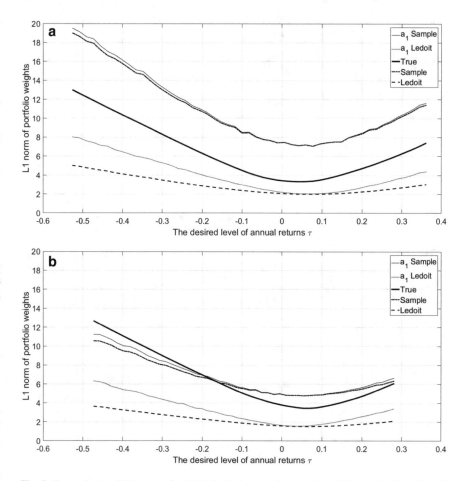

Fig. 9 Dependence of L1 norm of weights $\|\mathbf{w}\|$ of selected optimal portfolios on the desired level τ. (**a**) French market, N=108. (**b**) German market, N=150

More information can be obtained from the portfolio trajectories, formed by each point $(\hat{w}^{\top}\mathbb{E}(R), \mathbb{D})$ for every level τ. The trajectories are shown in Figure 10. An unexpected discovery here is that shrinkage estimator for means enables to achieve with higher expected returns portfolios from nearly the same frontier. Without these estimators, expected returns of selected portfolios were overrated because of estimation error in the mean vector, and thus the high desired levels τ were unattainable out-of-sample. This is the essence of the phenomenon of bias. Also, the figure demonstrates another evidence of advantages of Wolf–Ledoit matrix. It allows to reach portfolios from the frontier, which is significantly closer to the true efficient frontier.

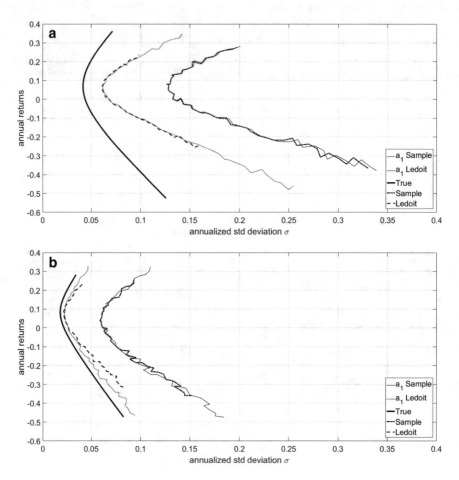

Fig. 10 Portfolio trajectories. (**a**) French market, N=108. (**b**) German market, N=150

To make the essence of the phenomenon more evident, we compare the results of selected portfolios with and without shrinkage estimator for means (see Figure 11) for different desired levels of annual return $\tau = 0.1, 0.15$, and 0.2. Quadrangles denote true optimal portfolios, triangles denote points on the portfolio trajectory for **Ledoit**, and circles for a_1 **Ledoit**. The figure demonstrates that due to poor estimation of means without shrinkage estimators, selected portfolios tend to have annual lower annual returns. This feature appears for both markets, but for German it is more apparent.

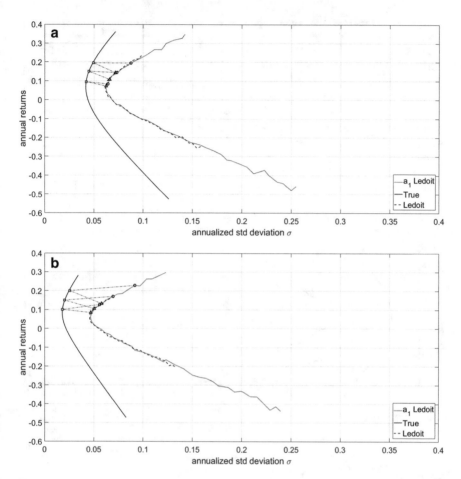

Fig. 11 Portfolio trajectories with compared results for different τ levels. (**a**) French market, N=108. (**b**) German market, N=150

3.4 Student t Distribution

Although shrinkage estimators might help to partially deal with estimation error in mean vectors, the results presented in the previous section show the performance of selected portfolios with generated normal data. So, it is necessary to test the estimators with samples generated from the other distributions. In this section, the same experiments were conducted, but the data were drawn from multinomial Student t distribution with 3 degrees of freedom and the same parameters in the generator. The empirical results are displayed in Figures 12 and 13.

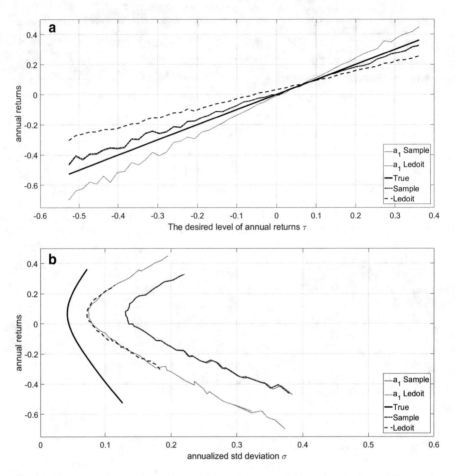

Fig. 12 French market. (**a**) Dependence of annual returns of selected optimal portfolios on the desired level τ with shrinkage means. (**b**) Portfolio trajectories

As it can be seen from the figures, the results with Students t distribution in the generator are roughly the same. This distribution has heavier tails than Normal distribution, but results did not change.

4   Future Research

Data driven approach became more and more popular in practical optimization problems [3]. It is interesting to continue the present research on the large scale portfolio optimization using this approach.

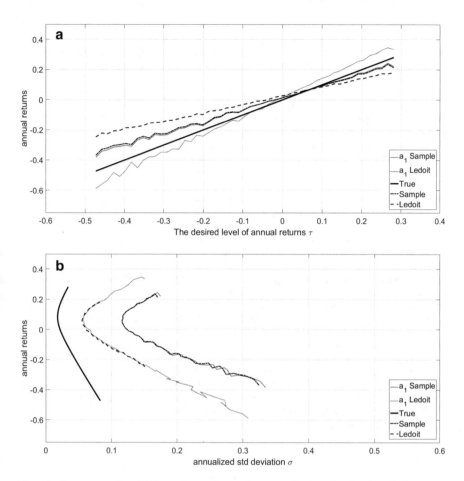

Fig. 13 German market. (**a**) Dependence of annual returns of selected optimal portfolios on the desired level τ with shrinkage means. (**b**) Portfolio trajectories

Acknowledgements The Sections 1–2 of the paper were prepared within the framework of the Basic Research Program at the National Research University Higher School of Economics (HSE). The author Kalyagin V. is partially supported by RSF grant 14–41-00039 (Section 3).

References

1. Altman, A., Gondzio, J.: Regularized symmetric indefinite systems in interior point methods for linear and quadratic optimization. Optim. Methods Softw. (1999)
2. Artzner, P., Delbaen, F., Eber, J.M., Heath, D.: Coherent Measures of Risk, Mathematical Finance (2011); Beale, E.M.L.: On minimizing a convex function subject to linear inequalities. J. R. Stat. Soc. B **17**, 173–184 (1955)

3. Bertsimas, D., Gupta, V. Kallus, N.: Data-driven robust optimization. Math. Program. A **167**, 235–292 (2018). https://doi.org/10.1007/s10107-017-1125-8
4. Best, M.J., Grauer, R.R.: On the sensitivity of mean-variance-efficient portfolios to changes in asset means: some analytical and computational results. Rev. Financ. Stud. **4**(2), 315–342 (1991)
5. Chopra, V.K., Ziemba, W.T.: The effect of errors in means, variances, and covariances on optimal portfolio choice. J. Portf. Manag. **19**, 6–11 (1993)
6. Connor, G., Korajczyk, R.A.: The arbitrage pricing theory and multifactor models of asset returns. Finance Handbook, Chapter 4 (1995)
7. James, W., Stein, C.: Estimation with quadratic loss. In: Proc. Fourth Berkeley Symp. on Math. Statist. and Prob., vol. 1, pp. 361–379. Univ. of Calif. Press (1961)
8. Jorion, P.: Bayes-Stein estimation for portfolio analysis. J. Financ. Quant. Anal. **21**(3), 279–292 (1986)
9. Kolm, P.N., Tutuncu, R., Fabozzi, F.: 60 Years of portfolio optimization: Practical challenges and current trends. Eur. J. Oper. Res. **234**(2), 356–371 (2014)
10. Ledoit, O.: Portfolio Selection: Improved Covariance Matrix Estimation (1994)
11. Ledoit, O., Wolf, M.: Improved estimation of the covariance matrix of stock returns with an application to portfolio selection. J. Empir. Financ. (2003)
12. Ledoit, O., Wolf, M.: Honey, I Shrunk the Sample Covariance Matrix (2004)
13. Lyuu, Y.D.: Financial Engineering and Computation: Principles, Mathematics, Algorithms. Cambridge University Press, Cambridge (2002)
14. Markowitz, H.: Portfolio selection. J. Financ. **7**, 77–91 (1952)
15. Nocedal, J., Wright, S.J.: Numerical Optimization, 2nd edn. Springer Series in Operations Research, Chapter 18. Springer (2006)
16. Roll, R.: A critique of the asset pricing theory's test; part I: On past and potential testability of the theory. J. Financ. Econ. **4**, 129–176 (1977)
17. Sharpe, W.F.: A simplified model for portfolio analysis. Manage. Sci. **9**, 277–293 (1963)
18. Stein, C.: Inadmissibility of the usual estimator for the mean of a multivariate normal distribution. In: Proceedings of the Third Berkeley Symposium on Mathematical Statistics and Probability, Volume 1: Contributions to the Theory of Statistics, pp. 197–206. University of California Press, Berkeley, CA (1956)

Optimal Design of Smart Composites

Georgios K. Tairidis, Georgia Foutsitzi, and Georgios E. Stavroulakis

Abstract In the present chapter, optimal design problems related to smart composites are investigated. First, the mechanical properties of a smart composite can be tailored to meet required specifications. Beyond classical shape and layout optimization related to the layers of a composite, pointwise optimization leading to functionally graded composites or even topology optimization can be applied. A cantilever beam with two materials is briefly presented. Furthermore, the control subsystem has several parameters to be optimized: number and position of sensors and actuators, as well as the parameters of the controller. Here, some basic techniques regarding soft control based on fuzzy and neuro-fuzzy strategies are presented, along with optimization options and methods which can be used for the fine-tuning of the parameters of the system. The main concept of the present chapter is to provide stimuli to those who deal with design, optimization, and control issues on smart structures.

1 Introduction

The optimal design of smart composite materials and structures has attracted a lot of scientific interest during the last years, as the availability of technological tools has increased dramatically. The optimization process usually deals with the shape or the thickness of the material, the direction of fibers, etc., see among others [1–4].

Besides the high availability of the design tools, the availability of the control tools has increased as well. Modern control schemes based on fuzzy and adaptive

G. K. Tairidis · G. E. Stavroulakis (✉)
School of Production Engineering and Management, Technical University of Crete, Institute of Computational Mechanics and Optimization, Chania, Greece
e-mail: gestavr@dpem.tuc.gr

G. Foutsitzi
Department of Informatics and Telecommunication, University of Ioannina, Preveza, Greece
e-mail: gfoutsi@uoi.gr

© Springer Nature Switzerland AG 2019
I. C. Demetriou, P. M. Pardalos (eds.), *Approximation and Optimization*,
Springer Optimization and Its Applications 145,
https://doi.org/10.1007/978-3-030-12767-1_10

neuro-fuzzy methods are capable of solving hard problems with less or even without any information about the examined model. In this direction, several optimization techniques (e.g., ANFIS, genetic algorithms, etc.) can be used for the improvement of the characteristics of the control as well [5, 6]. These tools can be easily programmed and used within commercial simulation packages and/or programming languages like Simulink and MATLAB.

Another option, which attracts a lot of interest, on the optimal design of smart materials and structures, lies on the use of topology optimization tools [7]. These techniques lead to the design of optimal microstructures or simply give useful information for material, sensor, or actuator placement. An important aspect here is that topology optimization results can be now applied to real life problems, due to the availability of 3D printing and additive manufacturing technology. A simple example of gradually changing two-material composite beam is presented here.

Moreover, the basic concept of soft control based on fuzzy and neuro-fuzzy methods is presented as a stimulus for those who are interested in the optimal design of materials and structures. It is shown that the available tools for the optimal tuning of the control mechanisms exist, in order to achieve the maximum potential from every method. These tools include among others the use of artificial neural networks, genetic algorithms [5], particle swarm optimization [6, 8], etc.

In the present investigation, several numerical examples from these fields are presented and discussed. Firstly, a smart beam model with two different materials (aluminum and polypropylene) is studied. The need for this investigation occurs from the rapid development of 3D printing processes with two materials, focusing on the accurate construction of models with complicated geometry. More specifically, the dynamical characteristics of the proposed model are calculated using modal analysis techniques, in order to find the influence of these two materials in the structural modes. The analysis was carried out using the finite element method through the Comsol Multiphysics software. The results of the analysis were compared with a simple homogeneous aluminum beam.

Moreover, an example from a composite plate with adhesive layers is considered. In this case a coupled electromechanical model of a piezoelectric plate under the layerwise theory is considered. The presence of adhesive material enables the study of delamination phenomena. Besides the capabilities of static or dynamic analysis, several control tools can be used for vibration suppression. In the examples which are given herein, firstly a simple fuzzy controller, and subsequently an optimized one are presented. The optimization is based on genetic algorithms. Such algorithms are nature-inspired global optimization tools which simulate the evolution process of species, including humankind.

Finally, numerical examples from the control field are discussed in order to show not only the capabilities of control in structural design, but also the availability of optimization techniques and their synergy with the whole system in order to build a robust and efficient control procedure.

2 Numerical Solution of Multiphysics Problems

One of the most common and powerful methods for the discretization and the analysis of smart structures is the finite element method (FEM). After the discretization of the structure, modal analysis tools and time-stepping techniques can be used for the determination of the dynamic behavior of the system and its response to various excitations. Mechanical, electrical, thermal, or other continuum mechanics layers can be solved together, interacting between them, by using the same framework.

The need for solving difficult and/or large-scale problems has led to the development of approximation methods. This is due to the fact that various technical problems are described by equations which are hard to be solved, except if it comes for very simple, small-scale problems defined on simple areas. The finite element method is an important tool for the numerical analysis and solution of several problems in mechanics. It is a very reliable approximation method and with major advantage the application on a very wide range of different problems, which are described by differential equations, as are the equations of motion in dynamical systems. The major drawback of this method lies to the fact that it is very costly, in terms of computational cost, especially in complex problems with complicated geometry and in non-linear problems. However, this disadvantage is overcome from the rapid development of computers.

The first thing that is required for the application of the finite element method is the geometry of the structure, which is usually inserted by using a computer aided design (CAD) program in order to develop the final structural model in two or three dimensions. Once the model is constructed, it is discretized in small elements (see Figure 1) which are called finite elements and a grid or mesh is created.

For the development of a mesh of adequate quality, the designer should be able to consider the main parameters of the model [9]. First of all, one should know approximately the optimal number of elements. However, besides the total number of elements, the geometry of the elements plays significant role as well to the mesh quality. The main criteria which are taken into account for the development of high quality elements are the element skewness and the element aspect ratio.

With the term skewness is denoted the error between the optimal and the real size of each element. The desired value of element skewness should not exceed 0.25. The element skewness is given by the formula:

$$\text{Skewness} = (\text{Optimal Cell Size} - \text{Cell Size})/\text{Optimal Size}.$$

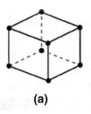   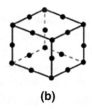   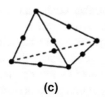

(a)   (b)   (c)

Fig. 1 Different types of finite elements

Fig. 2 Aspect ratio for
different element geometries

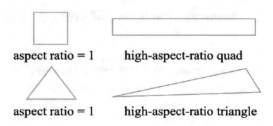

The element aspect ratio is the ratio between the larger and the smaller side of the element (triangle, quad, etc.) as seen in Figure 2. Ideally, this ratio should be equal or near 1. In this case the shape of the element is uniform (e.g., an isosceles triangle, a square, etc.). When designing a mesh of a structure, values of aspect ratio should be above the critical limit (e.g., above 0.7).

Then the designer should insert all the required data (e.g., loadings, boundary conditions, etc.) for the description of the considered boundary value problem and choose or program an appropriate solver. The commercial packages which are used for this purpose are usually called pre-processors. Once the data for solving are ready, they are usually inserted in the program or package which will go through the analysis. These programs are called solvers and they use numerical methods for the solving of the arising system of equations. Finally, once the solving step is fulfilled, a post-processor, i.e., a suitable program for the evaluation of the results, should be used. However, the designer may use only one software package (all-in-one) for every single step of the method. In the results which are presented in the present chapter, Solidworks is used for the design of model (CAD) and COMSOL Multiphysics is used for analysis, i.e., the application of the finite element method. It is worth mentioning that COMSOL software provides an integrated designing system; thus, it can be used as an all-in-one software for both the design and the analysis.

Furthermore, in order to avoid the complexity of a general three-dimensional finite element model and have flexibility to apply size-reduction techniques and coupling with various controllers, the authors have developed in-house finite element codes based on simplified mechanical bending theories for multilayered beams and structures. These techniques are outlined in the next section.

3 Structural Mathematical Model for Composites

A laminated composite plate with integrated piezoelectric sensors and actuators and adhesive layers between them is considered (Figure 3). The mid-plane of the first layer is set to coincide with the origin of the z-axis. For simplicity of the notation, all the non-adhesive layers will be considered piezoelectric. Elastic layers are obtained by setting their piezoelectric coefficients to zero. In order to be able to model delamination between the layers, a composite discrete-layer (layerwise) plate

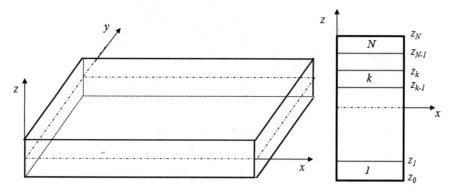

Fig. 3 Geometry of the smart piezocomposite plate

theory is employed. The formulation results in a coupled finite element model with mechanical (displacement) and electrical (potentials of piezoelectric layer) degrees of freedom. In addition the investigation is restricted to orthotropic layers, while extension to more general models is straightforward. Specially orthotropic layers are considered.

3.1 Displacement and Strains of the Non-adhesive Layers

The displacement field is defined according to a partial layerwise theory where the in-plane displacements $u_1^{(i)}$ and $u_2^{(i)}$ of the i-th layer are assumed linear through thickness, whereas transverse ones $u_3^{(i)}$ are supposed constant

$$u_1^{(i)}(x, y, z, t) = u^{(i)}(x, y, t) + (z - \tilde{z}^{(i)})\theta_x^{(i)}(x, y, t)$$
$$u_2^{(i)}(x, y, z, t) = v^{(i)}(x, y, t) + (z - \tilde{z}^{(i)})\theta_y^{(i)}(x, y, t)$$
$$u_3^{(i)}(x, y, z, t) = w^{(i)}(x, y, t), \tag{1}$$

where $u^{(i)}$, $v^{(i)}$, and $w^{(i)}$ are the mid-plane deformations of the i-th layer, $\theta_x^{(i)}$ and $\theta_y^{(i)}$ are rotation angles of the normal to the mid-plane about the y and x axes, respectively, and $\tilde{z}^{(i)}$ is the thickness of the mid-plane of the i-th layer.

The in-plane strain of the i-th layer can be expressed by the

$$\{\varepsilon_b^{(i)}\} = \{\varepsilon_{xx}^{(i)}, \varepsilon_{yy}^{(i)}, \gamma_{xy}^{(i)}\}^T = \{\varepsilon_{b0}^{(i)}\} + (z - \tilde{z}^{(i)})\{k^{(i)}\}, \tag{2}$$

where

$$\{\varepsilon_{0b}^{(i)}\} = \left\{ \frac{\partial u^{(i)}}{\partial x}, \frac{\partial v^{(i)}}{\partial y}, \left(\frac{\partial u^{(i)}}{\partial y} + \frac{\partial v^{(i)}}{\partial x} \right) \right\}^T,$$

$$\{k^{(i)}\} = \left\{ \frac{\partial \theta_x^{(i)}}{\partial x}, \frac{\partial \theta_y^{(i)}}{\partial y}, \left(\frac{\partial \theta_x^{(i)}}{\partial y} + \frac{\partial \theta_y^{(i)}}{\partial x} \right) \right\}^T.$$

The transverse shear strains of the i-th layer are given by

$$\{\varepsilon_s^{(i)}\} = \{\gamma_{yz}^{(i)}, \gamma_{xy}^{(i)}\}^T = \{\epsilon_{0s}^{(i)}\}, \tag{3}$$

where

$$\{\varepsilon_{0s}\} = \left\{ \frac{\partial w^{(i)}}{\partial y} + \theta_y^{(i)}, \frac{\partial w^{(i)}}{\partial x} + \theta_x^{(i)} \right\}^T.$$

In the above equations, a superscript T denotes the transpose of a matrix.

3.2   Constitutive Equations of Piezoelectric Layer

The linear constitutive equations for the i-th piezoelectric layer with reference to its principal axes are given by

$$\{\hat{\sigma}^{(i)}\} = [\bar{Q}^{(i)}]\{\varepsilon^{(i)}\} - [\bar{e}^{(i)}]^T \{E^{(i)}\}$$
$$\{\hat{D}^{(i)}\} = [\bar{e}^{(i)}]\{\varepsilon^{(i)}\} + [\bar{\xi}^{(i)}]\{E^{(i)}\}, \tag{4}$$

where $\{\hat{\sigma}^{(i)}\}, \{\hat{\varepsilon}^{(i)}\}$, $\{\hat{D}^{(i)}\}$, and $\{E\}$ are stress, strain, electric displacement, and electric field vector of the i-th layer, respectively. $[\bar{Q}^{(i)}]$, $[\bar{e}^{(i)}]$, and $[\bar{\xi}^{(i)}]$ are plane-stress reduced stiffness coefficients, the piezoelectric coefficients, and the permittivity constant matrices of the i-th layer, respectively. The first part of Equation (4) describes the inverse piezoelectric effect, while the second part accounts the direct effect.

Next, we assume that the piezoelectric material exhibits orthorhombic 2 mm symmetry. After transforming Equation (4) to the global coordinate system (x, y, z) and separating the bending and shear related variables, the constitutive Equation (4) become

$$\{\sigma_b^{(i)}\} = [Q_b^{(i)}]\{\varepsilon_b^{(i)}\} - [e_b^{(i)}]^T \{E^{(i)}\}$$
$$\{\sigma_s^{(i)}\} = [Q_s^{(i)}]\{\varepsilon_s^{(i)}\} - [e_s^{(i)}]^T \{E^{(i)}\}$$
$$\{D^{(i)}\} = [e_b]\{\varepsilon_b^{(i)}\} + [e_s^{(i)}]\{\varepsilon_s^{(i)}\} + [\xi^{(i)}]\{E^{(i)}\}, \tag{5}$$

where $\{\sigma_b^{(i)}\} = \{\sigma_{xx}^{(i)}, \sigma_{yy}^{(i)}, \tau_{xy}^{(i)}\}^T$, $\{\sigma_s^{(i)}\} = \{\tau_{yz}^{(i)}, \tau_{xz}^{(i)}\}^T$, and

$$[Q_b^{(i)}] = \begin{bmatrix} Q_{11}^{(i)} & Q_{12}^{(i)} & Q_{16}^{(i)} \\ Q_{21}^{(i)} & Q_{22}^{(i)} & Q_{26}^{(i)} \\ Q_{16}^{(i)} & Q_{26}^{(i)} & Q_{66}^{(i)} \end{bmatrix}, \quad [Q_s^{(i)}] = \begin{bmatrix} Q_{44}^{(i)} & Q_{45}^{(i)} \\ Q_{45}^{(i)} & Q_{55}^{(i)} \end{bmatrix}$$

$$[e_b^{(i)}] = \begin{bmatrix} 0 & 0 & 0 \\ 0 & 0 & 0 \\ e_{31}^{(i)} & e_{32}^{(i)} & e_{36}^{(i)} \end{bmatrix}, \quad [e_s^{(i)}] = \begin{bmatrix} e_{14}^{(i)} & e_{15}^{(i)} \\ e_{25}^{(i)} & e_{25}^{(i)} \\ 0 & 0 \end{bmatrix}, \quad [\xi^{(i)}] = \begin{bmatrix} \xi_{11}^{(i)} & \xi_{12}^{(i)} & 0 \\ \xi_{21}^{(i)} & \xi_{22}^{(i)} & 0 \\ 0 & 0 & \xi_{33}^{(i)} \end{bmatrix}. \quad (6)$$

In Equation (6), $Q_{kl}^{(i)}$, $e_{kl}^{(i)}$ and $\xi_{kk}^{(i)}$ are the transformed reduced elastic, piezoelectric, and permittivity constants of the i-th layer, respectively. The detailed expressions for transformed material constants can be obtained from [10]. For non-piezoelectric layer the material constants $e_{kl}^{(i)}$ and $\xi_{kk}^{(i)}$ should be zero.

3.3 Electric Field

A constant transverse electrical field is assumed for the piezoelectric layers and the remaining in-plane components are supposed to vanish. For most of the typical piezoelectric laminate structures with relatively small thickness of the piezoelectric layers in comparison to the overall structure's thickness, the electric field inside the i-th piezoelectric layer can be expressed as

$$\{E^{(i)}\} = \begin{bmatrix} 0 & 0 & -\dfrac{1}{h^{(i)}} \end{bmatrix} \phi^{(i)} = [B_\phi^{(i)}]\phi^{(i)}, \quad (7)$$

where $h^{(i)}$ and $\phi^{(i)}$ are the thickness and the difference of electric potential of the i-th piezoelectric layer.

3.4 The Adhesive Layer

The adhesive layers between the host plate and the piezoelectrics are assumed to be very thin, and their deformation is linear. Only transverse normal stress $\sigma_z^{(a_i)}$ and strains $\epsilon_{zz}^{(a_i)}$, and in-plane shear stress $\tau_{xz}^{(a_i)}$, $\tau_{yz}^{(a_i)}$ and strains $\gamma_{xz}^{(a_i)}$, $\gamma_{yz}^{(a_i)}$ are taken into account. The in-plane stretching of the adhesive layer is neglected, since its stiffness in that direction is quite small. Also the adhesive layer is treated as an isotropic material. The linear deformation of the adhesive layers can be written in terms of the deformations of the adjacent structural layers [11, 12]

$$u_1^{(a_i)} = Z[u^{(i+1)} + \frac{h^{(i+1)}}{2}\theta_x^{(i+1)}] + [1 - Z][u^{(i)} - \frac{h^{(i)}}{2}\theta_x^{(i)}]$$

$$u_2^{(a_i)} = Z[v^{(i+1)} + \frac{h^{(i+1)}}{2}\theta_y^{(i+1)}] + [1 - Z][v^{(i)} - \frac{h^{(i)}}{2}\theta_y^{(i)}]$$

$$u_3^{(a_i)} = Zw^{(i+1)} + [1 - Z]w^{(i)}, \tag{8}$$

where $Z = \frac{1}{h^{(a_i)}}(z - \tilde{z}^{(i)} - \frac{h^{(i)}}{2})$.

The above equations can be written in more compact form as follows:

$$\{\bar{u}^{(i)}\} = \{u^{(i)}, v^{(i)}, w^{(i)}, \theta_x^{(i)}, \theta_y^{(i)}\}^T = [R_t^{(a_i)}(z)]\{\bar{u}^{(i+1)}\} + [R_b^{(a_i)}(z)]\{\bar{u}^{(i)}\},$$
$$\tag{9}$$

where

$$[R_t^{(a_i)}(z)] = \begin{bmatrix} Z & 0 & 0 & \frac{h^{(i+1)}}{2}Z & 0 \\ 0 & Z & 0 & 0 & \frac{h^{(i+1)}}{2}Z \\ 0 & 0 & Z & 0 & 0 \end{bmatrix},$$

$$[R_b^{(a_i)}(z)] = \frac{1}{h^{(a_i)}}\begin{bmatrix} 1-Z & 0 & 0 & -\frac{h^{(i)}}{2}Z & 0 \\ 0 & 1-Z & 0 & 0 & -\frac{h^{(i)}}{2}Z \\ 0 & 0 & 1-Z & 0 & 0 \end{bmatrix}.$$

Using the above relations and taking into account that the adherends are much thicker than the adhesive layers, the shear and peel strains of the adhesive layers can be written as (see [11, 12] for more details)

$$\varepsilon_{zz}^{(a_i)} = \frac{w^{(i+1)} - w^{(i)}}{h^{(a_i)}}$$

$$\gamma_{yz}^{(a_i)} = \frac{1}{h^{(a_i)}}[v^{(i+1)} - v^{(i)} + \frac{h^{(i)}}{2}\theta_y^{(i)} + \frac{h^{(i+1)}}{2}\theta_y^{(i+1)}]$$

$$\gamma_{xz}^{(a_i)} = \frac{1}{h^{(a_i)}}[u^{(i+1)} - u^{(i)} + \frac{h^{(i)}}{2}\theta_x^{(i)} + \frac{h^{(i+1)}}{2}\theta_x^{(i+1)}]. \tag{10}$$

Substituting the relations (8) into Equation (10) gives

$$\{\epsilon^{(a_i)}\} = \{\epsilon_{zz}^{(a_i)}, \gamma_{xz}^{(a_i)}, \gamma_{xz}^{(a_i)}\}^T = [L_t^{(a_i)}]\{\bar{u}^{(i+1)}\} + [L_b^{(a_i)}]\{\bar{u}^{(i)}\}, \tag{11}$$

where

$$[L_t^{(a_i)}] = \frac{1}{h^{(a_i)}} \begin{bmatrix} 0 & 0 & 1 & 0 & 0 \\ 0 & 1 & 0 & 0 & \frac{h^{(i+1)}}{2} \\ 1 & 0 & 0 & \frac{h^{(i+1)}}{2} & 0 \end{bmatrix}, \quad [L_b^{(a_i)}] = \frac{1}{h^{(a_i)}} \begin{bmatrix} 0 & 0 & -1 & 0 & 0 \\ 0 & -1 & 0 & 0 & \frac{h^{(i)}}{2} \\ -1 & 0 & 0 & \frac{h^{(i)}}{2} & 0 \end{bmatrix}.$$

The bending and shear stress in the adhesive layer can be written as

$$\{\sigma^{(a_i)}\} = [Q^{(a_i)}]\{\varepsilon^{(a_i)}\} \quad or \quad \begin{Bmatrix} \sigma_{zz}^{(a_i)} \\ \tau_{xz}^{(a_i)} \\ \tau_{xz}^{(a_i)} \end{Bmatrix} = \begin{bmatrix} E^{(a_i)} & 0 & 0 \\ 0 & G^{(a_i)} & 0 \\ 0 & 0 & G^{(a_i)} \end{bmatrix} \begin{Bmatrix} \varepsilon_{zz}^{(a_i)} \\ \gamma_{xz}^{(a_i)} \\ \gamma_{xz}^{(a_i)} \end{Bmatrix}.$$

$$(12)$$

3.5   Finite Element Formulation

In this present study, the three-layered plate model has been discretized using a twelve-nodded isoparametric quadrilateral Lagrangian element with five degrees of freedom (DOF) per node (see Figure 4).

The element is developed to include the adhesive layer flexibility. The generalized displacement vector is interpolated as

$$\{\bar{u}^{(i)}(x, y, t)\} \equiv \{u^{(i)}, v^{(i)}, w^{(i)}, \theta_x^{(i)}, \theta_y^{(i)}\}^T = [N_u]\{d^{(i)}\}_e = \sum_{j=1}^{4} (N_j[I]_{5\times5}\{d_j\}),$$

$$(13)$$

where $\{d^{(i)}\}_e = \{\{d_1^{(i)}\}^T, \{d_2^{(i)}\}^T, \{d_3^{(i)}\}^T, \{d_4^{(i)}\}^T\}^T$ and $\{d_j^{(i)}\} = \{u_j^{(i)}, v_j^{(i)}, w_j^{(i)}, \theta_{xj}^{(i)}, \theta_{yj}^{(i)}\}^T$, $j = 1, 2, 3, 4$, corresponding to the j-th node of the element, $N_j, j = 1, 2, 3, 4$, are bilinear isoparametric shape functions, and $[I]_{5\times5}$ is the unit matrix.

Fig. 4 Twelve-nodded isoparametric element with five degrees of freedom per node

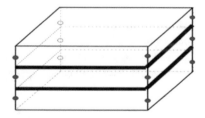

Substituting (13) into Equations (2) and (3) gives

$$\{\bar{\varepsilon}^{(i)}(x, y, t)\} = [B]\{d^{(i)}\} = \sum_{j=1}^{4} ([B_j]\{d_j^{(i)}\}) \tag{14}$$

or equivalent

$$\{\bar{\varepsilon}^{(i)}\} = \left\{ \begin{array}{c} \{\varepsilon_{b0}^{(i)}\} \\ \{k^{(i)}\} \\ \{\varepsilon_{s0}^{(i)}\} \end{array} \right\} = \left[\begin{array}{c} [B_b] \\ [B_k] \\ [B_s] \end{array} \right] \{d^{(i)}\} = \sum_{j=1}^{4} \left(\left[\begin{array}{c} [B_b]_j \\ [B_k]_j \\ [B_s]_j \end{array} \right] \{d_j^{(i)}\} \right),$$

where

$$[B_b]_j = \begin{bmatrix} \partial_x & 0 & 0 & 0 & 0 \\ 0 & \partial_y & 0 & 0 & 0 \\ \partial_y & \partial_x & 0 & 0 & 0 \end{bmatrix} N_j, \quad [B_k]_j = \begin{bmatrix} 0 & 0 & 0 & \partial_x & 0 \\ 0 & 0 & 0 & 0 & \partial_y \\ 0 & 0 & 0 & \partial_y & \partial_x \end{bmatrix} N_j, \quad [B_s]_i = \begin{bmatrix} 0 & 0 & \partial_x & 1 & 0 \\ 0 & 0 & \partial_y & 0 & 1 \end{bmatrix} N_j,$$

and $\partial_x = \frac{\partial}{\partial x}, \partial_y = \frac{\partial}{\partial y}$.

3.6 Variational Principle

This formulation will be based on the Hamilton variational principle in which the strain potential energy, kinetic energy, and work are considered for the entire structure. Since we are dealing with the piezoelectric continuum with adhesive interfaces, the Lagrangian and the virtual work are properly adapted to include the electrical and mechanical contributions as well as the contribution of the adhesive. The most general form of this variational principle is stated as

$$\int_{0}^{T} (\delta T - \delta U + \delta W) \, dt, \tag{15}$$

where T is the total kinetic energy, U is the total strain energy, and W is the work done by the loads.

The total kinetic energy of the system is the sum of the corresponding energies of individual layers and can be given by

$$T = \sum_{i=1}^{3} T^{(i)} + \sum_{i=1}^{2} T^{(a_i)}. \tag{16}$$

The kinetic energy of the i-th layer is given by

$$T^{(i)} = \frac{1}{2} \int_{V^{(i)}} \rho^{(i)} \left[\{\dot{u}_1^{(i)}\}^2 + \{\dot{u}_2^{(i)}\}^2 + \{\dot{u}_3^{(i)}\}^2 \right] dV, \tag{17}$$

where $\rho^{(i)}$ is the density of the i-th layer. Substituting the displacements relations (1), Equation (17) becomes

$$T^{(i)} = \frac{1}{2} \int_{V^{(i)}} \rho^{(i)} [(\dot{u}^{(i)})^2 + 2(z - \bar{z}^{(i)})\dot{u}^{(i)}\dot{\theta}_x^{(i)} + (\dot{v}^{(i)})^2 + 2(z - \bar{z}^{(i)})\dot{v}^{(i)}\dot{\theta}_y^{(i)} + (\dot{w}^{(i)})^2$$

$$+ (z - \bar{z}^{(i)})^2(\dot{\theta}_x^{(i)})^2 + (z - \bar{z}^{(i)})^2(\dot{\theta}_y^{(i)})^2] dV$$

$$= \frac{1}{2} \int_{V^{(i)}} \rho^{(i)} \begin{Bmatrix} u \\ v \\ w \\ \theta_x \\ \theta_y \end{Bmatrix}^T \begin{bmatrix} 1 & 0 & 0 & (z - \bar{z}^{(i)}) & 0 \\ 0 & 1 & 0 & 0 & (z - \bar{z}^{(i)}) \\ 0 & 0 & 1 & 0 & 0 \\ (z - \bar{z}^{(i)}) & 0 & 0 & (z - \bar{z}^{(i)})^2 & 0 \\ 0 & (z - \bar{z}^{(i)}) & 0 & 0 & (z - \bar{z}^{(i)})^2 \end{bmatrix} \begin{Bmatrix} u \\ v \\ w \\ \theta_x \\ \theta_y \end{Bmatrix} dV$$

$$= \frac{1}{2} \int_{V^{(i)}} \{\dot{u}^{(i)}\}^T [I^{(i)}(z)]\{\dot{u}^{(i)}\} dV. \tag{18}$$

Substituting Equation (13) in the relation (18), one obtains

$$T^{(i)} = \frac{1}{2}\{\dot{d}^{(i)}\}^T \int_{V^{(i)}} [N]^T [I^{(i)}(z)][N]dV\{\dot{d}^{(i)}\} \equiv \frac{1}{2}\{\dot{d}^{(i)}\}^T [M^{(i)}]\{\dot{d}^{(i)}\}. \tag{19}$$

The kinetic energy of the a_i-th adhesive layer is given by

$$T^{(a_i)} = \frac{1}{2} \int_{V^{(a_i)}} \rho^{(a_i)} \left[\{\dot{u}_1^{(a_i)}\}^2 + \{\dot{u}_2^{(a_i)}\}^2 + \{\dot{u}_3^{(a_i)}\}^2 \right] dV$$

$$= \frac{1}{2} \int_{V^{(a_i)}} \{\dot{u}^{(a_i)}\}^T [I^{(a_i)}(z)]\{\dot{u}^{(a_i)}\} dV, \tag{20}$$

where $[I^{(a_i)}(z)] = \rho^{(a_i)}[I]_{3 \times 3}$ and $V^{(a_i)}$ is the volume of the a_i-th adhesive layer. Using the relations (9) and (13), the above relation takes the form

$$T^{(a_i)} = \frac{1}{2}\{\dot{d}^{(i+1)}\}^T \int_{V^{(a_i)}} [N]^T [R_t^{(a_i)}]^T [I^{(a_i)}(z)][R_t^{(a_i)}][N]dV\{\dot{d}^{(i+1)}\}$$

$$+ \frac{1}{2}\{\dot{d}^{(i+1)}\}^T \int_{V^{(a_i)}} [N]^T [R_t^{(a_i)}]^T [I^{(a_i)}(z)][R_b^{(a_i)}][N]dV\{\dot{d}^{(i)}\}$$

$$+\frac{1}{2}\{\dot{d}^{(i)}\}^T \int_{V^{(a_i)}} [N]^T [R_b^{(a_i)}]^T [I^{(a_i)}(z)][R_t^{(a_i)}][N]\mathrm{d}V\{\dot{d}^{(i+1)}\}$$

$$+\frac{1}{2}\{\dot{d}^{(i)}\}^T \int_{V^{(a_i)}} [N]^T [R_b^{(a_i)}]^T [I^{(a_i)}(z)][R_b^{(a_i)}][N]\mathrm{d}V\{\dot{d}^{(i)}\}$$

$$\equiv \frac{1}{2}\{\dot{d}^{(i+1)}\}^T [M_{tt}^{(a_i)}]_e\{\dot{d}^{(i+1)}\} + \frac{1}{2}\{\dot{d}^{(i+1)}\}^T [M_{tb}^{(a_i)}]_e\{\dot{d}^{(i)}\}$$

$$+\frac{1}{2}\{\dot{d}^{(i)}\}^T [M_{bt}^{(a_i)}]_e\{\dot{d}^{(i+1)}\} + \frac{1}{2}\{\dot{d}^{(i)}\}^T [M_{bb}^{(a_i)}]_e\{\dot{d}^{(i)}\}. \tag{21}$$

The total strain energy of the system is represented as

$$U = \sum_{i=1}^{3} U^{(i)} + \sum_{i=1}^{2} U^{(a_i)}. \tag{22}$$

The strain energy of the i-th piezoelectric layer is given by

$$U^{(i)} = \frac{1}{2} \int_{V(i)} (\{\varepsilon_b^{(i)}\}^T\{\sigma_b^{(i)}\} + \{\varepsilon_s^{(i)}\}^T\{\sigma_s^{(i)}\})\,\mathrm{d}V$$

$$= \frac{1}{2} \int_{V(i)} (\{\varepsilon_{b0}^{(i)}\}^T[Q_b^{(i)}]\{\varepsilon_{b0}^{(i)}\} + \{\varepsilon_{b0}^{(i)}\}^T (z - \tilde{z}^{(i)})[Q_b^{(i)}]$$

$$+\{k^{(i)}\}^T (z - \tilde{z}^{(i)})\{\varepsilon_{b0}^{(i)}\} + \{k^{(i)}\}^T (z - \tilde{z}^{(i)})^2 [Q_b^{(i)}]\{k^{(i)}\}$$

$$+\{\varepsilon_{s0}^{(i)}\}^T [Q_s^{(i)}]\{\varepsilon_{s0}^{(i)}\} + \{k_s^{(i)}\}^T (z - \tilde{z}^{(i)})^2 [Q_s^{(i)}]\{\varepsilon_{s0}^{(i)}\}$$

$$-\{\varepsilon_{b0}^{(i)}\}^T [e_b^{(i)}]^T \{E^{(i)}\} - \{k^{(i)}\}^T (z - \tilde{z}^{(i)})[e_b^{(i)}]^T \{E^{(i)}\}$$

$$-\{\varepsilon_{s0}^{(i)}\}^T [e_s^{(i)}]^T \{E^{(i)}\})\,\mathrm{d}V$$

$$= \frac{1}{2} \int_{V(i)} \begin{bmatrix} \{\varepsilon_{b0}^{(i)}\} \\ \{k^{(i)}\} \\ \{\varepsilon_{s0}^{(i)}\} \end{bmatrix}^T \begin{bmatrix} [Q_b^{(i)}] & (z - \tilde{z}^{(i)})[Q_b^{(i)}] & 0 \\ (z - \tilde{z}^{(i)})[Q_b^{(i)}] & (z - \tilde{z}^{(i)})^2[Q_b^{(i)}] & 0 \\ 0 & 0 & [Q_s^{(i)}] \end{bmatrix} \begin{bmatrix} \{\varepsilon_{b0}^{(i)}\} \\ \{k^{(i)}\} \\ \{\varepsilon_{s0}^{(i)}\} \end{bmatrix} \mathrm{d}V$$

$$-\frac{1}{2} \int_{V(i)} \begin{bmatrix} \{\varepsilon_{b0}^{(i)}\} \\ \{k^{(i)}\} \\ \{\varepsilon_{s0}^{(i)}\} \end{bmatrix}^T \begin{bmatrix} [e_b^{(i)}]^T \\ (z - \tilde{z}^{(i)})[e_b^{(i)}]^T \\ [e_s^{(i)}]^T \end{bmatrix} \{E^{(i)}\}\mathrm{d}V$$

$$= \frac{1}{2} \int_{V(i)} (\{\bar{\varepsilon}^{(i)}\}^T [\mathcal{D}^{(i)}((z - \tilde{z}^{(i)}))]\{\bar{\varepsilon}^{(i)}\} - \{\bar{\varepsilon}^{(i)}\}^T [\mathcal{E}^{(i)}(z)]\{E^{(i)}\})\mathrm{d}V. \tag{23}$$

Substituting for $\{\epsilon_{b0}^{(i)}\}$, $\{k^{(i)}\}$, $\{\epsilon_{s0}^{(i)}\}$, and $\{E^{(i)}\}$ in Equation (23), $U^{(i)}$ can be written as

$$U^{(i)} = \frac{1}{2}\{d^{(i)}\}_e^T [K_{uu}^{(i)}]_e \{d^{(i)}\}_e - \frac{1}{2}\{d\}_e^T [K_{u\phi}^{(i)}]_e \{\phi^{(i)}\}_e, \qquad (24)$$

where

$$[K_{uu}^{(i)}]_e = \left[\int_{V^{(i)}} ([B]^T [\mathcal{D}^{(i)}(z)][B]dV \right], \quad [K_{u\phi}^{(i)}]_e = \left[\int_{V^{(i)}} [B]^T [\mathcal{E}^{(i)}(z)][B_\phi]^{(i)}dV \right].$$

The strain energy of the a_i-th adhesive layer is given by

$$U^{(a_i)} = \frac{1}{2} \int_{V^{(a_i)}} \{\epsilon^{(a_i)}\}^T \{\sigma^{(a_i)}\}dV = \frac{1}{2} \int_{V^{(a_i)}} \{\epsilon^{(a_i)}\}^T \{[Q^{(a_i)}]\}\{\epsilon^{(a_i)}\}dV. \qquad (25)$$

With the aid of Equations (11) and (13), the above equation takes the form

$$U^{(a_i)} = \frac{1}{2}\{d^{(i+1)}\}^T \int_{V^{(a_i)}} [N]^T [L_t^{(a_i)}]^T [Q^{(a_i)}][L_t^{(a_i)}][N]dV \{d^{(i+1)}\}$$

$$+ \frac{1}{2}\{d^{(i+1)}\}^T \int_{V^{(a_i)}} [N]^T [L_t^{(a_i)}]^T [Q^{(a_i)}][L_b^{(a_i)}][N]dV \{d^{(i)}\}$$

$$+ \frac{1}{2}\{d^{(i)}\}^T \int_{V^{(a_i)}} [N]^T [L_b^{(a_i)}]^T [Q^{(a_i)}][L_t^{(a_i)}][N]dV \{d^{(i+1)}\}$$

$$+ \frac{1}{2}\{d^{(i)}\}^T \int_{V^{(a_i)}} [N]^T [L_b^{(a_i)}]^T [Q^{(a_i)}][L_b^{(a_i)}][N]dV \{d^{(i)}\}$$

$$\equiv \frac{1}{2}\{d^{(i+1)}\}^T [K_{tt}^{(a_i)}]_e \{d^{(i+1)}\} + \frac{1}{2}\{d^{(i+1)}\}^T [K_{tb}^{(a_i)}]_e \{d^{(i)}\}$$

$$+ \frac{1}{2}\{d^{(i)}\}^T [K_{bt}^{(a_i)}]_e \{d^{(i+1)}\} + \frac{1}{2}\{d^{(i)}\}^T [K_{bb}^{(a_i)}]_e \{d^{(i)}\}. \qquad (26)$$

The total work W is the sum of the work done by the electrical forces W_E and the work done by the mechanical forces W_m. Using constitutive relations, strain displacement, and electric field–electric potential relations, the element electrical energy can be written as

$$W_E^{(i)} = \frac{1}{2} \int\limits_{V^{(i)}} \{E^{(i)}\}^T \{D^{(i)}\} dV$$

$$= \frac{1}{2} \int\limits_{V^{(i)}} \{E^{(i)}\}^T ([e_b^{(i)}]\{\epsilon_b^{(i)}\} + [e_s^{(i)}]\{\epsilon_s^{(i)}\} + [\xi^{(i)}]\{E^{(i)}\}) dV$$

$$= \frac{1}{2} \{\phi^{(i)}\}^T \int\limits_{V^{(i)}} [B_\phi^{(i)}]^T [\xi][B] dV \{d^{(i)}\} + \frac{1}{2} \{\phi^{(i)}\}^T \int\limits_{V^{(i)}} [B_\phi^{(i)}]^T [\xi^{(i)}][B_\phi^{(i)}] dV \{\phi^{(i)}\}_e$$

$$= \frac{1}{2} \{\phi^{(i)}\}^T [K_{\phi u}^{(i)}]\{d^{(i)}\} + \frac{1}{2} \{\phi^{(i)}\}^T [K_{\phi\phi}^{(i)}]\{\phi^{(i)}\}. \tag{27}$$

The work done by the mechanical forces is given by

$$W_m^{(i)} = \{\bar{u}^{(i)}\}^T \{f_c^{(i)}\} + \int\limits_{S_1^{(i)}} \{\bar{u}^{(i)}\}^T \{f_s^{(i)}\} dS + \int\limits_V \{\bar{u}^{(i)}\}^T \{f_v^{(i)}\} dV^{(i)}$$

$$- \int\limits_{S_2^{(i)}} \{E^{(i)}\}^T \{f_\phi\} dS$$

$$= \{d^{(i)}\}_e^T [N]^T \{f_c^{(i)}\} + \{d^{(i)}\}_e^T \int\limits_{S_1} [N]^T \{f_s^{(i)}\} dS + \{d^{(i)}\}_e^T \int\limits_V [N]^T \{f_v^{(i)}\} dV$$

$$- \{\phi^{(i)}\}_e^T \int\limits_{S_2} [B_\phi^{(i)}]^T \{f_\phi^{(i)}\} dS$$

$$= \{d^{(i)}\}_e^T \{F_m^{(i)}\}_e + \{\phi^{(i)}\}_e^T \{F_\phi^{(i)}\}_e. \tag{28}$$

In Equation (28), $\{f_c^{(i)}\}$ denotes the concentrated forces intensity, $\{f_s^{(i)}\}$ and $\{f_v^{(i)}\}$ denote the surface and volume force intensity, respectively, and $\{f_\phi^{(i)}\}$ denotes the surface charge density. $S_1^{(i)}$ and $S_2^{(i)}$ are the surface areas where the mechanical forces and electrical charge are applied, respectively. $\{F_m^{(i)}\}_e$ are the applied mechanical forces of the layer i in an element and $\{F_\phi^{(i)}\}_e$ are the applied electrical charges of the layer i in an element.

3.7 Equations of Motion

Using Hamilton's principle (15) the resultant global finite element spatial model, governing the motion and electric charge equilibrium of each element, is given by

$$[M]_e \left\{ \ddot{d} \right\}_e + [K_{uu}]_e \{d\}_e + \left[K_{u\phi} \right]_e \{\phi\}_e = \{F_m\}_e$$

$$\left[K_{\phi u} \right]_e \{d\}_e + \left[K_{\phi\phi} \right]_e \{\phi\}_e = \{F_\phi\}_e , \tag{29}$$

where $\{d\}_e = \{ \{d^{(1)}\}_e^T, \{d^{(2)}\}_e^T, \{d^{(3)}\}_e^T \}$, $\{\phi\}_e = \{ \phi_e^{(1)}, \phi_e^{(2)}, \phi_e^{(3)} \}$,

$$[M]_e = \begin{bmatrix} [M^{(1)}]_e + [M_{bb}^{(a_1)}]_e & [M_{bt}^{(a_1)}]_e & 0 \\ [M_{tb}^{(a_1)}]_e & [M^{(2)}]_e + [M_{tt}^{(a_1)}]_e + [M_{bb}^{(a_2)}]_e & [M_{bt}^{(a_2)}]_e \\ 0 & [M_{tb}^{(a_2)}]_e & [M^{(3)}]_e + [M_{tt}^{(a_2)}]_e \end{bmatrix}$$

$$[K]_e = \begin{bmatrix} [K^{(1)}]_e + [K_{bb}^{(a_1)}]_e & [K_{bt}^{(a_1)}]_e & 0 \\ [K_{tb}^{(a_1)}]_e & [K^{(2)}]_e + [K_{tt}^{(a_1)}]_e + [K_{bb}^{(a_2)}]_e & [K_{bt}^{(a_2)}]_e \\ 0 & [K_{tb}^{(a_2)}]_e & [K^{(3)}]_e + [K_{tt}^{(a_2)}]_e \end{bmatrix}$$

$$[K_{(u\phi)}]_e = \begin{bmatrix} K_{(u\phi)}^{(1)} & 0 & 0 \\ 0 & K_{(u\phi)}^{(2)} & 0 \\ 0 & 0 & K_{(u\phi)}^{(3)} \end{bmatrix}, \quad [K_{(\phi\phi)}]_e = \begin{bmatrix} K_{(\phi\phi)}^{(1)} & 0 & 0 \\ 0 & K_{(\phi\phi)}^{(2)} & 0 \\ 0 & 0 & K_{(\phi\phi)}^{(3)} \end{bmatrix}$$

$$\{F_m\}_e = \left\{ \{F_m^{(1)}\}_e^T, \{F_m^{(2)}\}_e^T, \{F_m^{(3)}\}_e^T \right\}^T, \{F_\phi\}_e = \left\{ \{F_\phi^{(1)}\}_e^T, \{F_\phi^{(2)}\}_e^T, \{F_\phi^{(3)}\}_e^T \right\}^T.$$

Following the routine of the assembly procedure, the global equations for the smart composite plate can be obtained.

The formulation is general in the sense that it can model laminated composite structures with arbitrary boundary conditions. The robustness of this formulation is that each element of each layer can be made of any material and if it is piezoelectric, then setting the appropriate (electrical) boundary conditions will allow it to act as a sensor or actuator.

4   Modal and Dynamic Analysis

For the calculation of the dynamical properties of the structure, modal analysis tools were used. For a linear elastic system with given mass and stiffness, its dynamic identity is quantified by the eigenfrequencies and the corresponding eigenmodes. The eigenvalues are calculated from the mass and stiffness matrices according to

$$\det(\mathbf{K} - \lambda \mathbf{M}) = 0. \tag{30}$$

For every value which is calculated by Equation (15), corresponds a displacement vector $a_{i,j}$.

This vector is called eigenvector of the eigenvalue problem, and provides the eigenvectors ϕ_i of the structures after normalization:

$$\phi_i = c_i \alpha_{i,j}, \tag{31}$$

where c_i is an arbitrary constant.

The eigenvectors are very important when designing a dynamic system, because they can help the designer of the system to monitor the several modes (ways) of vibration. Through these vectors, one can obtain useful information for the deformation and the displacement of the examined system. In technological applications, usually a small number of only the first eigenvectors are necessary in the sense that they capture the most of the kinetic energy of the system, and thus these eigenfrequencies together with the corresponding modes are calculated and considered for the design. This is due to the fact that, on the one hand, the excitation forces in nature excite only the first (low) eigenfrequencies of a structural system, and on the other hand, the inertial forces have smaller influence on the dynamic behavior of the system, as the excitation frequency increases. Hence, the modes with lower contribution to the response of the system, according to engineering experience, i.e., those with minor energy participation, are usually not taken into account. This assumption is important in order to simplify the examined problem.

The alteration to the eigenmodes and/or the eigenvalues of a structural system can be used as an accurate alert for the modification of the structural characteristics, in order to decide which is the best geometry and/or material and/or combination of materials for a specified criterion which is set during the designing process. Moreover, the modification of the eigen characteristics of a structure can provide useful information for failures or damages in the context of structural health monitoring (SHM).

For the dynamic analysis, i.e., for the integration of the equations of motion, several numerical methods have been proposed in the literature. In general, these algorithms can be used for the calculation of the numerical values approximating the response of the dynamical system within a definite interval of time and are very common in numerical analysis. In structural dynamics, ordinary differential equations are involved for the description of motion. The methods which are used in order to find the solutions of these equations are called ODE solvers. In this direction, several algorithms are available. The most common methods include among other the Euler method, the Runge–Kutta method, the Newmark-β method, the Houbolt method, etc.

The Houbolt numerical integration method which is very popular in engineering applications involving low frequencies is applied as follows. At first, one needs to choose properly the two Houbolt factors. For constant acceleration assumption within every time step, which is usually sufficient in structural dynamics, these parameters are set to

$$\beta = 0.25, \gamma = 0.5. \tag{32}$$

Also, the integration time t and the time step Δt need to be chosen by the designer of the system.

Integration constants are given as

$$c_1 = \frac{1}{\beta \Delta t^2}, c_2 = \frac{1}{\beta \Delta t}, c_3 = \frac{1}{2\beta}, c_4 = \frac{\gamma}{\beta \Delta t}, c_5 = \frac{\gamma}{\beta}, c_6 = \Delta t \frac{\gamma}{2\beta} - 1. \quad (33)$$

The Houbolt integration algorithm pseudocode can be written as follows:

Step 1: Initialization of variables
$u, \dot{u}, \ddot{u}, F_m, M, C, \beta, \gamma, c_1, c_2, c_3, c_4, c_5, c_6$

Step 2: Calculation of intermediate matrix $F* : * = +c_1 M + c_4 C$
Inversion of matrix $* : F* = (K*)^T$
Start of loop for t_0 to t_f

Step 3: Calculation of intermediate matrix $P*$
Calculation of difference of loadings: $dF_m = F_m(t + 1) - F_m(t)$
Calculation of difference of control force z
Aggregation to the quantity dF_m: $dF_m = dF_m + z$
Calculation of matrix $P*$ using mass matrix M and damping matrix C of the system:
$P* = dF_m + M[c_2\dot{u}(t) + c_3\ddot{u}(t)] + C[c_5\dot{u}(t) + c_6\ddot{u}(t)]$

Step 4: Calculation of response step du
$du = F * P*$

Step 5: Solution for the next step $(t + t)$
Calculation of acceleration: $\ddot{u}(t + 1) = \ddot{u}(t) + c_1 du - c_2\dot{u}(t) - c_3\ddot{u}(t)$
Calculation of velocity: $\dot{u}(t + 1) = \dot{u}(t) + c_4 du - c_5\dot{u}(t) - c_6\ddot{u}(t)$
Calculation of displacement: $u(t + 1) = u(t) + du$

end for

5   Structural Control

5.1   Fuzzy Control in General

The classical fuzzy control systems, known also as fuzzy rule-based systems, are based on the fuzzy inference techniques, which in turn are based on fuzzy sets. Fuzzy inference provides a framework for quantitative usage of linguistic rules involving vague (fuzzy) parameters, a task which characterizes logical creatures like humans. A fuzzy inference system consists of a database of IF–THEN rules, a set of membership functions, a decision-making unit (inference process), a fuzzification interface, and a defuzzification interface. The operation of the inference system goes as follows. The explicit inputs are converted into fuzzy via fuzzification. Then the set of rules is drafted, which together with the data forms the knowledge database.

Subsequently, the decision is made by implication, and the fuzzy output arises. Finally, this value is defuzzified.

The systematic process which utilizes the verbal rules for the decision making is called aggregation. The verbal rules describe the relation between the fuzzy variables, i.e., the inputs and outputs, using logical operators. The aggregation takes into account the intersection of the involved sets when the AND operator is used, while in the case of the OR operator it uses the union of the fuzzy sets. These methods are also known as minimum (min) and maximum (max) method, respectively.

Fuzzification is an important process in fuzzy theory, as it converts an explicit numerical quantity into a fuzzy one, which is represented by the membership functions. The process is based on the recognition of the uncertainty which exists in explicit quantities. On practical applications it is possible for errors to occur with a consequent reduction of data accuracy. This reduction of precision can also be represented by the membership functions. The definition or fine-tuning of the membership functions can be done either intuitively or by using algorithms and logical processes. The most popular methods include intuition, inference, rank ordering, angular fuzzy sets, neural networks (in adaptive neuro-fuzzy systems), genetic algorithms (in optimized fuzzy systems), inductive reasoning, etc.

On the other hand, defuzzification is the conversion of fuzzy outputs into explicit values. This process is necessary as the value of outputs must be accurate, especially when the fuzzy system is used as a controller, where the fuzzy outputs are not useful for further processing. For the defuzzification of fuzzy output functions, one can consider several methods (see Figure 5).

The most commonly used methods include the maximum membership principle, the centroid, the bisector, the middle or mean of maximum (MOM), the smallest of maximum (SOM), the largest of maximum (LOM), the center of sums, the center of largest area, the weighted average, etc. The choice of the appropriate defuzzification

Fig. 5 An example of defuzzification methods (http://www.mathworks.com/ help/examples/fuzzy_ featured/defuzzdm_04.png)

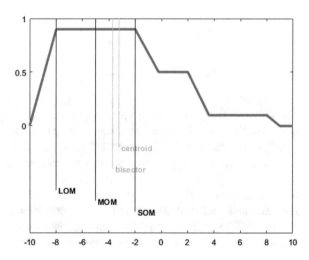

method is often a subjective process and depends on the data and/or the requirements of each problem. It is noteworthy that two different methods can give completely different results. It is also possible for the results of two or more methods to be identical. For example, if the final surface is triangular, the result of methods SOM, MOM, and LOM will be identical (the top of the triangle).

The two main methods of fuzzy inference are the Mamdani method and the Sugeno method. Other known methods are the Inference of Larsen and the Tsukamoto method. For the investigations considered in the present dissertation, two Mamdani-type fuzzy controllers and one Sugeno-type neuro-fuzzy controller were developed. The first method, which is the most widespread and will be considered herein, was introduced by Mamdani and Assilian in 1975 [13]. The main advantages of the Mamdani method include the fact that it is an intuitive method, it is widely accepted, and it adapts well to real problems. Moreover, it is a relatively simple method which works well even in complex models, without sacrificing accuracy.

The basic steps of the implementation process of the Mamdani method are:

1. Fuzzification of inputs using the membership functions,
2. Fuzzification of output using the membership functions,
3. Definition of verbal rules of fuzzy system,
4. Evaluation of the rules,
5. Calculation of system outputs, and
6. Defuzzification.

5.2   Development of a Simple Fuzzy Controller

In this section, a fuzzy controller is designed for vibration suppression of smart structures, taking into account the electromechanical formulation of a structure with piezoelectrics. The Mamdani inference is considered for the controller which consists of two inputs, the electric potential and the electric current, and one output, an electric signal which can be used for control. The membership functions of the variables of the controller are of triangular and trapezoidal form, for both inputs and the output (see Figures 6, 7, and 8).

It is worth mentioning that the range of the output, i.e., of the control signal of the actuator, is set to $[-200, 200]$ as the piezoelectric materials used for the analysis can produce voltage which is up to 200 V.

The set of rules is written based on the pendulum logic and is shown in Table 1 and in Figure 9. All rules have weights equal to unity and are connected using the AND operator.

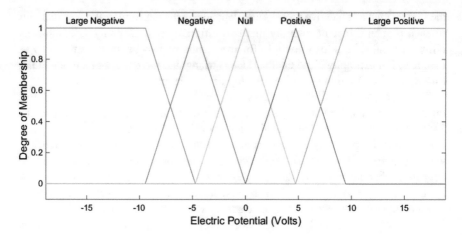

Fig. 6 Membership function of the electric potential (input 1)

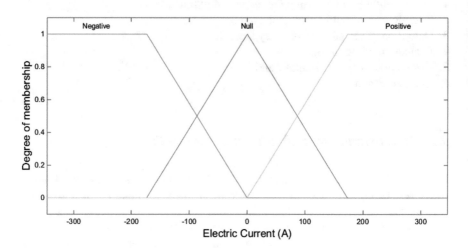

Fig. 7 Membership function of the electric current (input 2)

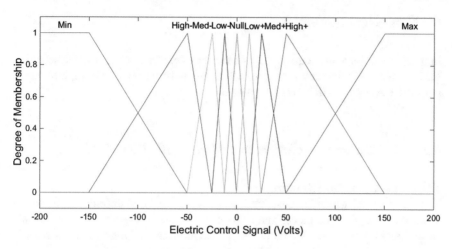

Fig. 8 Membership function of the electric control signal (output)

Table 1 Fuzzy inference rules for the electrical system (e.g., if electric potential is far positive and electric current is positive, then the electric control signal is max)

Electric potential / Electric current	Far positive	Close positive	Equilibrium	Close negative	Far negative
Positive	Max	Med+	Low+	Nul	Low−
Nul	Med+	Low+	Null	Low−	Med−
Negative	High+	Null	Low−	Med−	Min

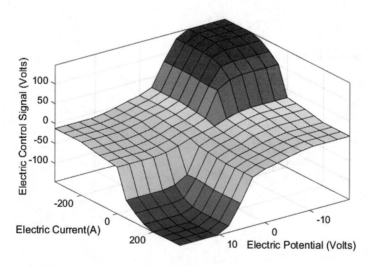

Fig. 9 Graphic representation of the fuzzy rules

6 Numerical Examples

In order to reinforce the acquaintance of the above contexts of the discrete scientific fields of structural dynamics, system modelling, and control, a set of representative numerical examples will be presented below.

6.1 An Example of a Structural Model with Two Materials

The Geometry of the Structure
The beam model which is considered here consists of three different parts, which however, act as a single piece. The contact surface is considered as a curve which converges to the symmetry axis of the beam from the left to the right, as seen in Figures 10 and 11. The dimensions are 70 mm × 5 mm × 6 mm.

For the specific problem which is presented here, two different materials were used. More specifically, for the upper and lower surface an aluminum (6061 T6) was used, while for the core of the beam a propylene material (3120 MU 5) was

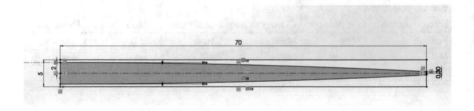

Fig. 10 Dimensions of the composite beam

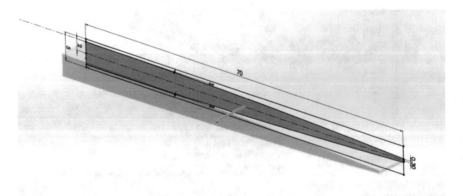

Fig. 11 The mid-plane surface of the beam

Table 2 Material properties

Material	Young's modulus	Poisson's ratio	Density
Aluminum	68.9 GPa	0.33	2.70 g/cc
Polypropylene	1.18 GPa	0.42	0.902 g/cc

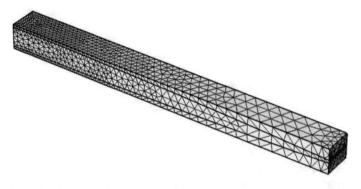

Fig. 12 The final mesh of the structure

Table 3 Eigenfrequencies for the different formulations

Eigenfreq. (Hz)	$L = 0.8$	$L = 1.6$	$L = 2.4$	$L = 3.2$	$L = 4.0$
1	0.58048	0.60188	0.62662	0.65562	0.69026
2	0.69221	0.71884	0.74977	0.78617	0.82978
3	3.5875	3.6166	3.676	3.7831	3.9667
4	4.1417	4.1537	4.1846	4.2093	4.1863
5	4.4462	4.4082	4.358	4.2887	4.2172

considered. The selection of these materials was based on the desired properties of the structure in terms of elasticity, and more specifically due to their much different Young's modulus, which will provide the desired behavior. The properties for the selected aluminum and propylene materials are shown in Table 2. The mesh of the structure is done by using tetrahedral elements as shown in Figure 12.

The total number of elements in the mesh is 10158 elements with average quality above 0.7.

For the analysis, five different sweep parameters were considered. Namely, the parameters $L = 0.8$, $L = 1.6$, $L = 2.4$, $L = 3.2$, and $L = 4.0$ were used. The results for the first five eigenfrequencies are given in Table 3.

The maximum displacement at each eigenmode for the several models is shown in Figure 13. Note that only the translational displacements along the y and z axis are taken into account. The case of a common aluminum rod is also presented for comparison.

In this example, the modal analysis has shown that the eigenfrequencies in the case of the simple aluminum rod are almost two-times bigger, compared with the

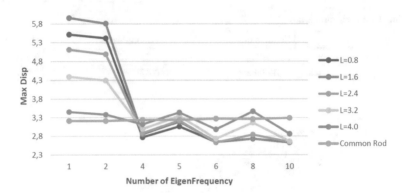

Fig. 13 Graphical comparison of the displacements for the simple and the composite beams

respective frequencies of the composite structures with two materials, both the elastic and the propylene material.

Moreover, it is shown that the insertion of the second material (propylene) the dynamic characteristics of the structure were improved, i.e., the eigenfrequencies were severely reduced. Such a composite structure would be suitable in several applications, and especially when high values of frequency occur (wind loadings, acoustics, etc.). In low frequency problems (e.g., seismic analysis problems) the simple elastic beam presents better behavior.

6.2 Fuzzy Control of Smart Plates in the Presence of Delamination

In this example, the capabilities of fuzzy control on smart composite plates under delamination conditions are examined [11, 14–16]. To be able to describe delamination phenomena, one may consider a layerwise model with adhesive material, as the one described in Section 2. The plate model considered for this investigation is shown in Figure 14.

The material properties of the composite structure are given in Table 4. Further information can be obtained from [17].

The stacking sequence of the plate is [0/±45/90]. The dimensions of the structure are 100 mm × 100 mm × 1.3 mm. 10 × 10 isoparametric elements are used for the analysis and realistic boundary conditions (RBC) are chosen, yielding to a system of 100 elements. Each finite element consists of 12 nodes and every node has 5 degrees of freedom (three translational and two rotational). Therefore, the structure has 363 nodes and 1815 degrees of freedom. Note that the upper layer is used as actuator, the middle layer represents the elastic core of the beam (graphite/epoxy), while the lower layer is used as sensor. One can observe that the numbering of nodes starts from the lower layer and is continuing to the other two.

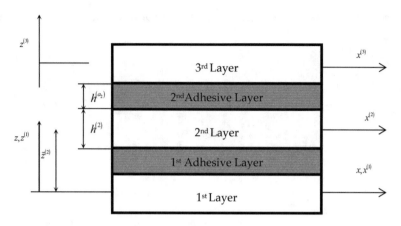

Fig. 14 A three-layered composite plate with adhesive materials

Table 4 Material properties of the composite plate with piezoelectrics and adhesive material

Property	Graphite/epoxy	Adhesive	Density
E_1 (GPa)	130	1.78	59
E_2 (GPa)	9.6	1.78	59
G_{12} (GPa)	4.8	–	–
G_{23} (GPa)	3.2	–	–
ν_{12}	0.31	0.3	0.34
ρ (kg/m^3)	1570	1050	7400
$d_{31} = d_{32}$ (m/V)	–	–	-260×10^{-12}
Ply thickness (mm)	0.1	0.05	0.2

It is also worth mentioning that the use of RBC is fundamental for the accurate prediction of the parameters of the closed loop system. This is due to the fact that piezoelectric components are able to actuate more effectively. In this context, fuzzy control is used for the vibration suppression of the multilayered plate both before and after the appearance of delamination. The delamination of the piezoelectric patch may occur either to the upper or the lower layer, which means that either the sensor or the actuator can peel off. The objective is the construction of robust controllers, which could be able to keep functioning under failure conditions, even if they are initially set without considering such phenomena.

6.2.1 Fuzzy Control on the Non-delaminated Coupled Electromechanical Model

In this paragraph, a structure without delamination of piezoelectric layers is investigated. In contrast to the results of the mechanical model, one can observe that the oscillation reduction in this case is not so satisfactory. This comes as a result of

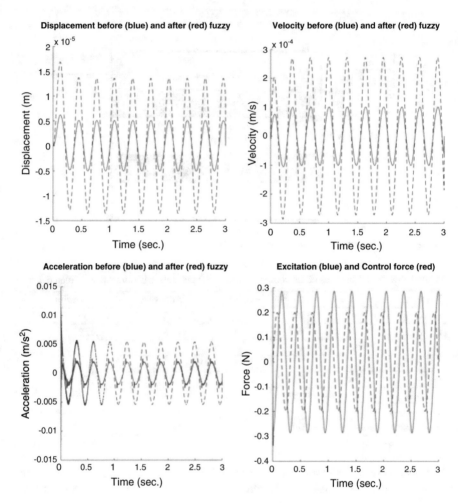

Fig. 15 Displacement, velocity, acceleration, and forces for the system without delamination (coupled electromechanical model)

the conversion of the displacement and velocity, in electrical potential and electric current, respectively, and the indirect transfer of the information on the mechanical model through the sensor and actuator. With blue color is denoted the vibration prior the application of fuzzy control, while the red color is used for the response after the application of the control. As for the forces, with blue and red color are denoted the external loading and the control force, respectively.

From the results shown in Figure 15, a satisfactory decrease of the oscillations in terms of displacement and velocity is achieved. One disadvantage is that the form of the acceleration is little rough, which makes the control quite unsatisfactory.

Fig. 16 Discretization of
membership functions for
optimization of categories

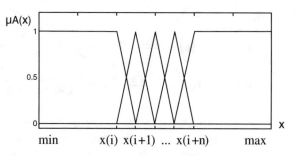

6.2.2 Optimization of Fuzzy Control with Genetic Algorithms

To improve the results obtained from the simple fuzzy controller, a fine-tuning
of its parameters is made. The main purpose of the optimization process is the
reduction of the displacement field and the improvement of the rough character of
the acceleration diagram. For this purpose, a genetic algorithm is implemented and
used to fine-tune the membership functions of the fuzzy variables.

The genetic algorithm which is used here is based on the algorithm proposed by
Mihalewicz [18] and is programmed in the MATLAB environment. The membership functions of the electric potential and the electric current were discretized in i,
j, or k points for each variable as shown in Figure 16. Due to symmetry reasons,
the zero point has been kept constant at the center of the functions.

The optimization problem consists of the maximization of the objective function
which is the percentage of oscillation reduction not only in terms of displacement,
but in terms of acceleration as well, as given in Equation (34). The problem is subject
to linear inequality constraints which are given in Equation (35). It is noted that the
maximum reduction is translated in minimum oscillations of the structure.

Maximize:

$$percentage = 0.4 \cdot percentage(u) + 0.6 \cdot percentage(\ddot{u}) \qquad (34)$$

Subject to:

$$
\begin{aligned}
x(i) &< x(i+1) && \text{if} \quad el.potential \geq 0 \\
x(i) &> x(i+1) && \text{if} \quad el.potential < 0 \\
x(j) &< x(j+1) && \text{if} \quad el.current \leq 0 \\
x(j) &> x(j+1) && \text{if} \quad el.current < 0,
\end{aligned}
\qquad (35)
$$

where

$$percentage(u) = \frac{(max\_displ\_before\_control) - (max\_displ\_after\_control)}{(max\_displ\_before\_control)}$$

$$percentage(\ddot{u}) = \frac{(max\_acc\_before\_control) - (max\_acc\_after\_control)}{(max\_acc\_before\_control)}$$

$$
\begin{aligned}
&max\_displ\_before\_control = max(u\_before) + |min(u\_before)| \\
&max\_displ\_after\_control = max(u\_after) + |min(u\_after)| \\
&max\_acc\_before\_control = max(\ddot{u}\_before) + |min(\ddot{u}\_before)| \\
&max\_acc\_after\_control = max(\ddot{u}\_after) + |min(\ddot{u}\_after)| \\
&0 < x(i) < 1, \quad i = 1, 4 \\
&0 < x(j) < 1, \quad j = 1, 2.
\end{aligned}
\tag{36}
$$

At this investigation, the population size was set to 50 members, that is, 50 different possible solutions. The maximum number of generations was set to 100. The initial population was chosen by a stochastic process with respect to the design variable's bounds. Regarding the selection, the Tournament Selection method was used, with the parameter q for the selection set to 2 members of the population. For the genetic operators of crossover and mutation, the random crossover B and the random non-uniform mutation were chosen with probabilities of 0.8 and 0.1, respectively. The results of the optimization process for the fuzzy inputs, that is, the electric potential and the electric current, are presented in Figures 17 and 18, respectively.

The results of the application of the optimized controller to the system without delamination are shown in Figure 19. The blue color denotes the vibration before the application of control, while with red color is denoted the response after the

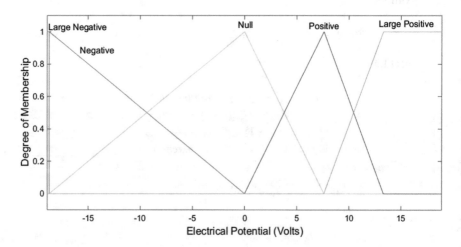

Fig. 17 Optimized membership function of the electric potential

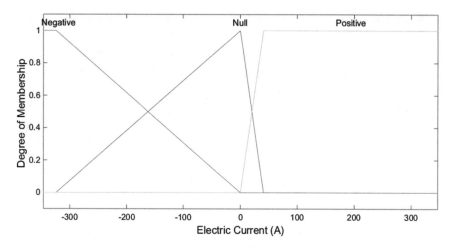

Fig. 18 Optimized membership function of the electric current

application of the fuzzy control. As for the forces, with blue color is denoted the external loading, while the control force is shown with red color.

From the results above, a slight improvement of the suppression in terms of displacement, velocity, and acceleration is noticed. In addition, the vibrations are smoother. This optimized controller is used for the reduction of the oscillations of the structure (considering the coupled electromechanical model) in the presence of delamination as shown in the following paragraph.

6.2.3 Fuzzy Control on the Delaminated, Coupled Electromechanical Model

A relatively significant partial delamination of 50% of the actuator is presented in this paragraph. The results are shown in Figure 20. The vibration prior and after the application of fuzzy control is denoted with blue and red color, respectively.

In this case, one can observe that the response of the control in terms of displacement, velocity, and acceleration is noticeably worse compared to the ones of the non-delaminated case; however, the reduction achieved remains satisfactory. This fact is very important for the robustness of control, as it proves that the fuzzy controller remains quite efficient, even under quite large amounts of delamination.

In Table 5, the percentage of the reduction of displacement, velocity, and acceleration for the investigations of the electrical model are presented in detail.

More information and investigations about different delamination cases can be found in [11].

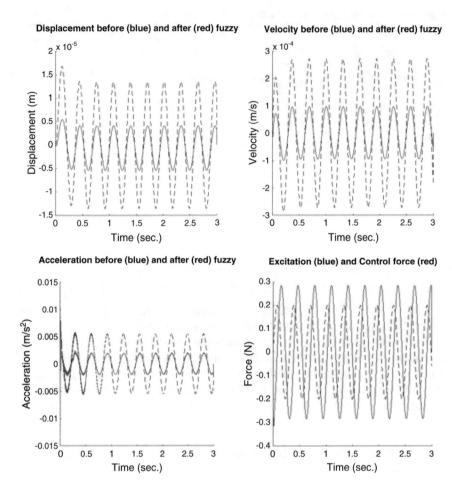

Fig. 19 Displacement, velocity, acceleration, and forces for the system without delamination and optimized membership functions (coupled electromechanical model)

7   Conclusions

From the numerical examples, which were presented in detail in this chapter, one may exclude some useful conclusions. First of all, flexible tools exist for the design of composite structures, that is, the piezoelectric materials and/or other smart materials (e.g., polypropylene plastics) for the alteration of the characteristics of the smart structure in order to achieve the desired properties, as, for example, the passive damping. The second, let say, step includes the optimization tools for the design, such as shape optimization as described above, or even the tools and principles of topology optimization.

The insertion of an external control mechanism, based, for example, on the principles of fuzzy logic, can be considered as the next step of the total designing

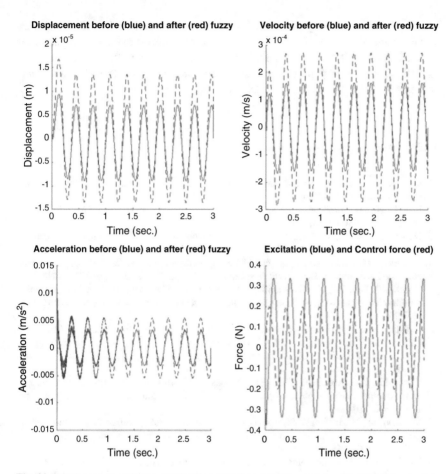

Fig. 20 Displacement, velocity, acceleration, and forces for the system with 50% delamination and optimized MF of fuzzy variables (coupled electromechanical model)

Table 5 Percentages of reduction of displacement, velocity, and acceleration

	0% delamination		50% delamination
Percentage of reduction	No optimization	Memb. Fcn. optimization	Memb. Fcn. optimization
Displacement	62.97%	69.9%	39.88%
Velocity	62.94%	64.62%	40.61%
Acceleration	53.77%	60.10%	33.71%

process. The results of fuzzy control can be evaluated as very satisfactory in terms of displacement; however, the suppression in terms of velocity and acceleration was below expectations. Thus, the characteristics of control are optimized in order to achieve optimum performance. From the investigations presented, it is clear that a genetic algorithm is a very powerful optimization tool which can be used to optimize and fine-tune fuzzy controllers. The oscillations of the smart plate were significantly reduced, not only in terms of displacement and velocity, but in terms of acceleration as well, which is a very important achievement. Moreover, the results were not only successful in terms of oscillations reduction, but smooth as well.

References

1. Muc, A., Kedzio, P.: Optimal design of smart laminated composite structures. Mater. Manuf. Processes. **25**(4), 272–280 (2010)
2. Irschik, H.: A review on static and dynamic shape control of structures by piezoelectric actuation. Eng. Struct. **24**, 5–11 (2002)
3. Onoda, J., Hanawa, Y.: Actuator placement optimization by genetic and improved simulated annealing algorithms. AIAA J. **31**, 1167–1169 (1993)
4. Koconis, D.B., Kollar, L.P., Springer, G.S.: Shape control of composite plates and shells with embedded actuators. II. Desired shape specified. J. Compos. Mater. **28**, 459–482 (1994)
5. Tairidis, G., Foutsitzi, G., Koutsianitis, P., Stavroulakis, G.E.: Fine tuning of a fuzzy controller for vibration suppression of smart plates using genetic algorithms. Adv. Eng. Softw. **101**, 123–135 (2016)
6. Tairidis, G.K., Foutsitzi, G., Koutsianitis, P., Stavroulakis, G.E.: Fine tuning of fuzzy controllers for vibration suppression of smart plates using particle swarm optimization. In: Proceedings of the 8th GRACM International Congress on Computational Mechanics, Volos (2015)
7. Kaminakis, N., Drosopoulos, G., Stavroulakis, G.E.: Design and verification of auxetic microstructures using topology optimization and homogenization. Arch. Appl. Mech. **85**, 1289–1306 (2015)
8. Marinaki, M., Marinakis, Y., Stavroulakis, G.E.: Fuzzy control optimized by a multi-objective particle swarm optimization algorithm for vibration suppression of smart structures. Struct. Multidisc. Optim. **43**, 29–42 (2011)
9. Wördenweber, B.: Finite element mesh generation. Comput. Aided Des. **16**, 285–291 (1984)
10. Reddy, N.J.: Mechanics of Laminated Composite Plates: Theory and Analysis. CRC Press, New York (1997)
11. Koutsianitis, P., Tairidis, G.K., Drosopoulos, G.A., Foutsitzi G.A., Stavroulakis, G.E.: Effectiveness of optimized fuzzy controllers on partially delaminated piezocomposites. Acta Mech. **228**, 1373–1392 (2017)
12. Erturk, C.L., Tekinalp, O.: A layer wise approach to piezoelectric plates accounting for adhesive and delaminated regions. Compos. Struct. **83**, 279–296 (2005)
13. Mamdani, E.H., Assilian, S.: An experiment in linguistic synthesis with a fuzzy logic controller. Int. J. Man. Mach. Stud. **7**, 1–13 (1975)
14. Tairidis, G.K., Foutsitzi, G.A., Koutsianitis, P., Drosopoulos, G., Stavroulakis, G.E.: Design and testing of fuzzy controllers on smart structures in the presence of delamination. In: Tsompanakis, Y., Kruis, J., Topping, B.H. (eds.) Proceedings of the Fourth International Conference on Soft Computing Technology in Civil, Structural and Environmental Engineering, Paper 41, Civil-Comp Press, Stirlingshire (2015)
15. Koutsianitis, P., Drosopoulos, G., Tairidis, G.K., Stavroulakis, G.E.: Optimally tuned fuzzy control for smart, possibly damaged piezocomposites. In: 3rd Polish Congress of Mechanics and 21st Computer Methods in Mechanics. CRC Press, Gdansk (2015)

16. Koutsianitis, P., Moutsopoulou, A., Drosopoulos, G., Tairidis, G.K., Foutsitzi, G., Stavroulakis, G.E.: Optimal control tuning in smart structures with delaminations. In: Hersonisos, Crete: ECCOMAS VII European Congress on Computational Methods in Applied Sciences and Engineering (2016)
17. Han, J.H., Lee, I.: Analysis of composite plates with piezoelectric actuators for vibration control using layerwise displacement theory. Compos. B **29**, 621–632 (1998)
18. Mihalewicz, Z.: Genetic Algorithms + Data Structures = Evolution Programs, 3rd edn. Springer, Berlin (1996)

Tax Evasion as an Optimal Solution to a Partially Observable Markov Decision Process

Paraskevi Papadopoulou and Dimitrios Hristu-Varsakelis

Abstract Motivated by the persistent phenomenon of tax evasion and the challenge of tax collection during economic crises, we explore the behavior of a risk-neutral self-interested firm that may engage in tax evasion to maximize its profits. The firm evolves in a tax system which includes many of "standard" features such as audits, penalties and occasional tax amnesties, and may be uncertain as to its tax status (not knowing, for example, whether a tax amnesty may be imminent). We show that the firm's dynamics can be expressed via a partially observable Markov decision process and use that model to compute the firm's optimal behavior and expected long-term discounted rewards in a variety of scenarios of practical interest. Going beyond previous work, we are able to investigate the effect of "leaks" or "pre-announcements" of any tax amnesties on the firm's behavior (and thus on tax revenues). We also compute the effect on firm behavior of any extensions of the statute of limitations within which the firm's tax filings can be audited, and show that such extensions can be a significant deterrent against tax evasion.

1 Introduction

In recent years, there has been increasing interest in the study of optimization and optimal control problems in the area of taxation and tax policy [5, 6, 11, 19, 20]. This activity is motivated in part by the recent global financial crisis which brought the problem of tax revenue collection squarely into the fore, but also by the availability of computational power which allows researchers to delve in areas where analytical results are scarce. This work explores an optimization problem related to tax evasion, a persistent phenomenon with which most countries grapple to some extent. In particular, we seek to determine the actions of a self-interested, risk-neutral tax

P. Papadopoulou · D. Hristu-Varsakelis (✉)
Computational Methods and Operations Research Laboratory, Department of Applied Informatics, University of Macedonia, Thessaloniki, Greece
e-mail: mai16036@uom.edu.gr; dcv@uom.gr

© Springer Nature Switzerland AG 2019
I. C. Demetriou, P. M. Pardalos (eds.), *Approximation and Optimization*, Springer Optimization and Its Applications 145, https://doi.org/10.1007/978-3-030-12767-1_11

entity (we will use the term "firm") which may engage in tax evasion in order to maximize its long-term profits. The firm will be allowed to evolve dynamically in a tax system which includes many of the features commonly used, including a tax rate, penalties for concealing profits, random audits (where the firm's past tax statements can be checked for a number of years into the past), and occasional optional tax amnesties which the government may offer but which the firm has no advance knowledge of. In this context, which we will make precise shortly, we are interested in (i) determining the firm's optimal behavior and expected long-term discounted rewards, (ii) finding out whether the firm can profit by reducing its uncertainty with respect to upcoming amnesties (e.g., from government announcements or "leaks" to the press), and (iii) quantifying the effect on maximal firm revenues (and thus on government revenues) of possible increases to the statute of limitations within which the firm's past actions can be audited.

Prior work related to optimal taxation and tax-evasion modeling includes early approaches, such as [1] which examined tax evasion as a portfolio allocation problem, and subsequent improvements (e.g., [4, 18, 33]). One disadvantage of these and later analytical approaches was that, in order to remain tractable, they often took on a macroscopic viewpoint, and could not express taxpayer heterogeneity nor could they fully capture the dynamics of tax evasion [21]. In recent years, there has been interest in modeling taxpayer entities at a finer-grained level in order to study their year-to-year evolution through the tax system [11]. These dynamics may include the random transitions in a firm's tax status (e.g., being subject to a surprise audit, or being included in an amnesty program) or the changing preferences of firms, viewed as interacting agents. In some cases, however, the richness of these models comes at a price in terms of complexity and requires a computational, rather than analytic, approach [9, 10].

The work in [11], on which this paper builds, examined a Markov-based model of the firm's evolution whose state corresponded to its yearly tax status (i.e., being audited or not, being able to expunge previous tax records through amnesty, etc.). That work showed that if the firm is risk-neutral, then its optimal policy can be computed via dynamic programming, and produced maps of the space of tax parameters identifying those that would remove the incentive for the firm to evade taxation. One limitation of [11] was the fact that the firm knew each year whether the government intended to activate a tax amnesty, and therefore could take advantage of that knowledge as it decided its future course. In practical settings (e.g., Greece, which [11] and the present work use as a case study) this may occur when the government creates expectations either through official announcements or press leaks. Under normal circumstances, however, the firm cannot be assumed to have information on upcoming amnesties. It is thus crucial to develop models that take into account the resulting uncertainty from the point of view of the firm as to its true tax status. Doing so will allow us to explore the effects of that uncertainty on the firm's actions with respect to tax evasion. At the same time, it is important for policy makers to know what are the consequences, if any (in terms of revenue), of information leaks.

This paper's contribution is twofold. With respect to describing the firm's behavior vis-a-vis tax evasion, we propose a model which is structurally more parsimonious and yet more realistic than its predecessors, taking into account the fact that the firm has incomplete knowledge of its tax status. Our model, in the form of a POMDP, will allow us to approximate the firm's optimal policy given the parameters of the tax environment, to investigate whether it is important for the government to be careful about the information it releases on possible tax amnesties, and to quantify the effect (in terms of revenue) of keeping taxpayers "in the dark" regarding upcoming amnesties. The proposed model is used to identify the combinations of tax penalties and audit probabilities that lead to honest behavior, and to quantify the impact that an increase in the statute of limitations on tax audits would have on tax evasion, an aspect which—to our knowledge—has not been sufficiently explored in the context of relevant dynamical models.

The remainder of this paper is structured as follows: In Section 2 we propose a POMDP model that captures the firm's evolution in the tax system. The firm seeks to maximize its discounted long-term expected profit, taking into account the rules imposed by the tax system and its awareness (or lack thereof) of any imminent opportunity for amnesty that the government may provide. Section 3 discusses the solutions obtained from our model and examines the impact of uncertainty on the firm's decisions and the circumstances under which it is useful for the firm to know the government's intentions in advance. In the same section we explore the effect of extending the statute of limitations on auditing and how it affects tax evasion.

2   A POMDP Model of Tax Evasion

We proceed to construct a mathematical description of the firm's time evolution in the tax system. We will consider a generic tax system similar to that in [11] (see that work for a fuller description), which includes a fixed tax rate on profits, random audits by the tax authorities, and tax penalties for underreporting income. For the sake of concreteness, we will use Greece as a case study [30] when it comes to selecting specific values for the various tax parameters, although the discussion applies to a much broader setting. During an audit, a firm's tax statements can be scrutinized for up to 5 years in the past, meaning that any tax evasion beyond that horizon goes unpunished. Audits tend to focus on firms that have not been audited for three or 4 years running, and thus have tax records which are about to pass beyond the statute of limitations. Finally, the Greek tax system occasionally offers a kind of optional tax amnesty [2], termed "closure," in which a firm may pay a one-time fee for excluding past tax statements from a possible audit. The use of tax amnesties is not unusual, with hundreds of cases documented across many countries, including the USA [25], India [7], and Russia [3]. In Greece, the closure option mentioned above was used during 1998–2006 [13, 14] and was again considered more recently [15]. The availability of the closure option gives an incentive for firms who evade taxation to pay *some* amount in taxes where they might otherwise pay none, as it increases the chance of an audit for those who do not participate. However, it has been shown [11] to encourage tax evasion.

Operationally, at the end of each fiscal year the firm declares its net profit to the government, and also its intent to use the closure option (and pay the associated fees) if it becomes available. In an important—and realistic—departure from previous models [11] the firm does not know in advance whether the option is to be made available or not for the current year; that information is released only after the tax filing deadline. Thus the firm must decide on its tax-evasion policy without knowing, for example, whether it will be able to expunge any imminent tax misdeeds by availing itself of the option. This introduces uncertainty as to the firm's tax status and gives rise to a POMDP [8, 17] which we describe next.

2.1   The Firm's State and Action Sets

Using the notation from [11], we will let $s_k \in S$ be the tax status of representative firm in year $k = 0, 1, 2, \ldots$ with

$$S = \{V_1, \ldots, V_5, O_1, \ldots, O_5, N_1, \ldots, N_5\}, \qquad (1)$$

where

- V_i: the firm is being audited for the last $i = 1, \ldots, 5$ tax filings,
- O_i: the government is making the closure option available to the firm, whose last audit or closure option usage occurred $i = 1, \ldots, 5$ years ago, and
- N_i: the firm is not being audited nor has a closure option available, and its last audit or closure option usage happened $i = 1, \ldots, 5$ years ago.

For the sake of notational convenience, we will sometimes refer to the elements of S by integer, in their order of their appearance, i.e., $V_1 \rightarrow 1$, $V_2 \rightarrow 2$, $\ldots$, $N_5 \rightarrow 15$. We note that S contains five "copies" of each type of tax status (V,O,N) corresponding to the 5-year statute of limitations on tax evasion. Of course, S (and the discussion that follows) could be appropriately generalized to model a longer, length L time-window on which the government has the chance to detect tax evasion:

$$S = \{V_1, \ldots, V_L, O_1, \ldots, O_L, N_1, \ldots, N_L\}.$$

We will have more to say about this in Section 3.

We will let $A = \{1, 2\} \times [0, 1]$ be the firm's action set, where at year k the firm selects $a_k \in A$, $a_k = [v_k, u_k]^T$, with $v \in \{1, 2\}$ denoting the firm's decision to apply for usage of the closure option ($v_k = 1$) or to forgo the closure option ($v_k = 2$), and $u_k \in [0, 1]$ being the fraction of profits that the firm decides to conceal. Based on the above, the firm's state at time k will be the vector

$$x_k = [s_k, h_k^T]^T, \qquad (2)$$

where $s_k \in \mathcal{S}$, and $h_k \in [0, 1]^5$ will contain a history of the firm's latest five decisions with respect to tax evasion.

We note that our model differs from that of [11] in one important and practical point. In [11], the state vector included one additional element, corresponding to whether or not the closure option is available to the firm or not. In our case, the firm has no such knowledge; it can declare its wish (v_k) to avail itself of the option (should the government make it available after the tax filing deadline) but is forced to commit to a decision on tax evasion (u_k) in advance. Put in other words, without any information as to the government's intent to offer the closure option, the states O_i are indistinguishable to the firm from their N_i counterparts at the time the firm makes its decisions.

2.2 State Evolution

Based on the above discussion (see also [11]), the firm will evolve in $\mathcal{S} \times [0, 1]^5$ according to

$$x_{k+1} = Ax_k + Ba_k + n_k, \quad x(0) \text{ given}, \tag{3}$$

where

$$A = \begin{bmatrix} 0 \\ & H \end{bmatrix} \quad H = \begin{bmatrix} 0 & 1 & 0 & 0 & 0 \\ 0 & 0 & 1 & 0 & 0 \\ 0 & 0 & 0 & 1 & 0 \\ 0 & 0 & 0 & 0 & 1 \\ 0 & 0 & 0 & 0 & 0 \end{bmatrix}, \quad B = \begin{bmatrix} 0 & 0 \\ \vdots & \vdots \\ 0 & 0 \\ 0 & 1 \end{bmatrix}, \quad n_k = \begin{bmatrix} w_k \\ 0_{5 \times 1} \end{bmatrix}, \tag{4}$$

and the term $w_k \in \mathcal{S}$ is a random process whose transition probabilities depend on whether the firm applies for use the closure option or not, and whether the government ultimately decides to grant it at that particular year. Based on our earlier description of the tax system, we can represent the process driving w_k in graphical form using two transition diagrams, one for the case where $v_k = 1$ (the firm decides to use the closure option, Figure 1) and another when $v_k = 2$ when the firm declines the use of the option (Figure 2).

For example, let us assume that in Figure 1 the firm has the tax status O_2. This means that the firm will pay to exclude its last two tax statements from any audits, thus "cleaning its slate," and will now file its next tax statement—the only one subject to a possible audit next year. The firm's possible transitions are to O_1 (if the government does grant the option again) where the just-filed tax statement will be expunged; to V_1 where it will be audited; or to N_1, where the firm will avoid an audit but its tax statement will be kept in waiting, for possible future audits or closures. The transition diagram in Figure 2 (where the firm has declined the closure option) operates in a similar manner, with the firm never transitioning to an O_i state.

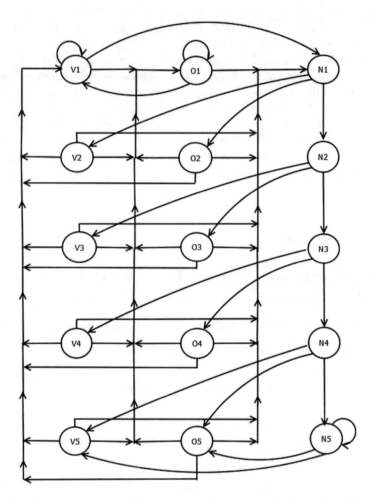

Fig. 1 Transition diagram when the firm asks to use the closure option. The values of the various transition probabilities are discussed in the text and in the Appendix

However, the transition probabilities from O_i states to audit states V_i will be higher compared to the previous case (see below and the Appendix for numerical values) to reflect the fact that once the government offers the closure option, any firm that opted out has a higher chance of being audited because its peers that opted in have now removed themselves from the audit pool.

2.2.1 Transition Probabilities

Based on the transition diagrams of Figures 1 and 2, the transition probabilities of the process w_k (and thus the firm's evolution from one tax status to another) are

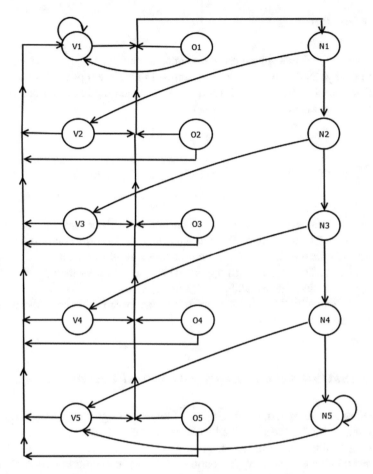

Fig. 2 Transition diagram when the firm decides not to use the closure option. The values of the various transition probabilities are discussed in the text and in the Appendix

$$Pr\,(s_{k+1} = i \,|\, s_k = j, a_k = (v_k, u_k)) = T_{ij}\,(v_k)\,, \quad i, j \in \{1, \ldots 15\}, \tag{5}$$

where T_{ij} is the (i, j)-th element of the transition matrix T given by

$$T\,(v_k) = \begin{cases} T_1 \text{ if } v_k = 1 \text{ (apply for closure)} \\ T_2 \text{ if } v_k = 2 \text{ (decline closure)} \end{cases} \tag{6}$$

and the T_1 and T_2 are shown in the Appendix.

2.3   Firm Rewards

Let Π denote the firm's annual profit (assumed to be constant), r the tax rate, β the annual penalty rate for any unpaid taxes applied in the event of an audit, and ℓ the cost of closure as a fraction of the firm's profits. Then the firm's reward is as per [11] (given here again for completeness):

$$
g(x, a) = g\left([s, h^T]^T, a\right) = \Pi \cdot \begin{cases} 1 - r + ru & s \in \{11, \ldots, 15\} \\ 1 - r + ru - \ell(s - 5) & s \in \{6, \ldots, 10\} \\ 1 - r + ru - r \sum_{i=1}^{s}[h]_{6-i} \\ -\frac{3}{5}\beta r \sum_{i=1}^{s} i[h]_{6-i} & s \in \{1, \ldots, 5\} \end{cases},
$$

(7)

where we have labeled elements of S by integer, and $[h]_i$ denotes the i-th element of the vector h. The top term on the right-hand side of (7) corresponds to the reward obtained if the firm is neither audited nor using the closure option and conceals an amount of Πu. The second term applies when the firm uses the option and thus pays ℓ per year since its last audit or closure. The last term is the firm's reward in the event of an audit, where its past history of tax evasion is used to calculate the back taxes and penalty owed.

2.4   Firm Observations, Belief, and Value Function

As we have already stated, the firm does not know if the government is to offer the closure option until *after* it has filed its taxes for the year. This means that—unless it is being audited—the firm is uncertain of its tax status (the first element of its state vector) which may be N_i (no option available) or O_i (option available if the firm is willing to pay for it). In practice, the firm may have some information from the press, government, or market sources, as to whether a new round of closure may be imminent. Of course, the information may not always be correct. Also, it is in the government's interest to know what effect any such "leaks" would have on firm behavior (and thus on tax revenues).

We will let $\hat{S}$ denote the set of observations that the firm makes of its tax status, with

$$
\hat{S} = \{\hat{V}_1, \ldots, \hat{V}_5, \hat{O}_1, \ldots, \hat{O}_5, \hat{N}_1, \ldots, \hat{N}_5\}.
$$

(8)

Thus, at time k, the firm observes $\hat{s}_k \in \hat{S}$, based on which it must decide on its course of action a_k. When the firm is in an audit state, it is of course aware of that situation and thus

$$
P\left(\hat{s}_k = \hat{V}_i | s_k = V_i\right) = 1,
$$

(9)

while in the case when the firm has an O_i or N_i status,

$$P\left(\hat{s}_k = \hat{O}_i | s_k = O_i\right) = z_O, \quad P\left(\hat{s}_k = \hat{N}_i | s_k = O_i\right) = 1 - z_O, \tag{10}$$

$$P\left(\hat{s}_k = \hat{N}_i | s_k = N_i\right) = z_N, \quad P\left(\hat{s}_k = \hat{O}_i | s_k = N_i\right) = 1 - z_N, \tag{11}$$

where z_O is the probability of correctly distinguishing an O_i state from its N_i counterpart, while z_N is the probability of correctly observing N_i versus O_i.

Given the state evolution equation (3) and observations $\hat{s}_k$ we can construct the firm's belief as a probability distribution over its states, which is to be updated with every new observation made by the firm [17]. It is important to note that the uncertainty in the firm's observations has to do solely with its tax status, i.e., the first element s_k of the state vector $x_k = [s_k, h_k^T]^T$. The rest of the state vector, h_k, is the firm's tax history which is of course always known to the firm. In light of this, we may define the observations in the entire state space to be

$$\hat{x}_k = [\hat{s}_k, h_k^T]^T,$$

and Equations (9)–(11) determine the observation probabilities of all states, so that it is sufficient to consider the belief $b(s)$ as a probability distribution on $\mathcal{S}$, instead of $b(x)$ on the entire state space. The firm's belief $b_k(s_k)$ after taking action a_k and observing $\hat{s}_{k+1}$ will then be updated to

$$b_{k+1}(s_{k+1}) = c \, P(\hat{s}_{k+1} | s_{k+1}) \sum_{s_k \in \mathcal{S}} Pr(s_{k+1} | s_k, a_k) b_k(s_k), \tag{12}$$

where c is a normalizing constant, and in (12) we have used the fact that observations depend only on the state the firm is in and not on its actions.

Given the firm's current belief, $b_k(s_k)$, as to its tax status, known tax history h_k, and action a_k, we can calculate its expected reward over the belief distribution, based on Equation (7):

$$R(b_k, h_k, a_k) = \sum_{s_k \in S} g\left([s_k, h_k^T]^T, a_k\right) b_k(s_k). \tag{13}$$

Assuming an infinite time horizon, the firm is then faced with the problem choosing its closure and tax-evasion decisions $a_k = \pi(b_k, h_k) = (v_k, u_k)$, so as to maximize, over the policy π, its discounted expected reward:

$$J^\pi(b_0, h_0) = \sum_{k=0}^{\infty} \gamma^k \mathop{\mathbb{E}}_{w_k} \left\{ g\left([s_k, h_k^T]^T, a_k\right) | b_0, h_0, \pi \right\} \tag{14}$$

subject to the dynamics (3) and observations (12), where $\gamma \in (0, 1]$ is a discount factor. The optimal value function J^* in Equation (14) obeys the well-known [8, 17] Bellman equation

$$J^*(b_k, h_k) = \max_{a_k} \left[R(b_k, h_k, a_k) + \gamma \sum_{\hat{s}_{k+1} \in \hat{\mathcal{S}}} P(\hat{s}_{k+1} | b_k, a_k) J^*(b_{k+1}, h_{k+1}) \right]$$
(15)

with

$$P(\hat{s} | b_k, a_k) = \sum_{s, s_k \in \mathcal{S}} P(\hat{s} | s) Pr(s | s_k, a_k) b_k(s_k).$$

2.5 Solving for the Firm'S Optimal Policy

Solving Equation (15) for the firm's optimal policy is generally difficult given (i) the fact that the number of states that the firm can occupy is not countable (recall that the firm's state includes its history of tax-evasion decisions, which are real numbers in the unit interval), and (ii) the difficulties associated with partial observability of the state. As is often the case when it comes to optimizing POMDPs, we will be able to make progress only by approximating the optimal value function, by means of an iterative process [24, 27].

Because we have assumed that the firm is risk-neutral, the reward function (7) is linear in the fraction of the profits to be concealed, u, making J^* linear as well. This implies that J^* will be maximized at the boundary of u's feasible region, meaning that we only need to consider the two extreme values $u = 0$ or $u = 1$ each year (i.e., be completely honest or conceal as much profit as possible—see [11] for a fuller discussion on this point). This yields a significant reduction in computational complexity, because it will be sufficient to consider $h_k \in \{0, 1\}^5$ and solve the Bellman equation only for a finite set of $|\mathcal{S}| \cdot 2^5 = 480$ states. This is to be compared to the 869 states in [11]. Of course, our model is more challenging computationally, because of the uncertainty in state observations.

To solve for the firm's optimal policy, we used the PERSEUS point-based value iteration algorithm [28]. Point-based algorithms became popular as methods of approximating POMDP policies [22, 23]. They rely on the fact that performing many fast approximate updates to a set of policy/value samples often results in a more useful value function than performing a few exact updates [16, 26]. The algorithm in [29] differs from other point-based algorithms in that at each iteration it "backs up" only a random subset of points in the firm's belief space. Doing so leads to computational savings so that the method can afford to use a larger number of samples compared to other point-based methods, and obtain better accuracy [29, 32]. In the next section we present a series of numerical experiments, where we optimize the firm's policy using [28], and discuss the results.

3 Results and Discussion

To begin with, we would like to determine the optimal policy and expected firm revenue depending on the availability of closure and the probability of the firm correctly observing its own state. We considered three cases with respect to the closure option: (i) available every fiscal year, (ii) never available, and (iii) given randomly, with a fixed probability of $q = 0.2$. With respect to the firm's observation probabilities, we considered the cases when the firm is (i) fully aware of the government's decision with respect to closure (i.e., it can correctly observe whether it is in a N state or an O state; this will allow us to validate our model against known solutions), or (ii) has no information on the government's intentions and must guess as to its state (i.e., $z_O = z_N = 0.5$). The remaining tax parameters were set as in [11] using Greece as a case study for the sake of being specific and to facilitate comparisons with previous work, namely, a tax rate of $r = 0.24$, penalty rate $\beta = 0.24$, cost of closure $\ell = 0.023$, and discount factor $\gamma = 0.971$ corresponding to a 3% annual rate of inflation.

3.1 Model Validation: The Case of "Perfect" Observations

In the case where the closure option is given randomly but the firm can make perfect observations of its state, the probability of a "correct" tax status observation in Equations (11)–(10) is $z_O = z_N = 1$. Then, the firm's optimization problem reduces to a Markov decision process (MDP) similar to that in [11] which can be solved easily via value iteration. Furthermore, the cases where the closure option is either always or never available also imply perfect observations because then the firm (who is not being audited) can safely conclude that it is in an O_i or N_i state, respectively. Table 1 shows the firm's optimal discounted expected reward in each case. As we can see, the reward is highest when the government offers the closure option with probability 1. In that case, the optimal policy was a constant $a = (1, 1)$, i.e., for the firm to always use the closure option every single year and to conceal all profit. In fact, from the point of view of government revenue, it is best to never offer the closure option. The firm's optimal policies agreed with the exact solutions obtained by [11], as did the firm's expected rewards (to within 0.11%).

Table 1 Comparison of the firm's discounted expected reward, under perfect state observation, with $r = 0.24$, $\beta = 0.24$, $\ell = 0.023$, and a 5% overall audit probability. Numbers are in % of the firm's annual profit Π, with a discount factor γ corresponding to a 3% annual rate of inflation

Closure option	Always available	Available with 20%	Never available
Exp. reward	3354.3	3307.1	3254.5

3.2 The Role of Uncertain Observations

Next, we investigate whether the firm's uncertainty as to its tax status may affect its tax-evasion behavior and, ultimately, its expected revenue. To this end, we must choose a specific probability for the closure option being offered at any year. We set $q = 1/5$, which is in some agreement with empirical data from Greece [11]; this rate corresponds to the government attempting to collect some revenue—in the form of closure fees—from firms which it does not have the resources to audit and whose tax filings are about to pass beyond the five year statute of limitations.

We solved for the firm's optimal policy and computed its expected revenue under: (i) perfect observations, where the firm always knows its tax status (equivalently, the government's decision to offer closure or not, $z_O = z_N = 1$), (ii) observations which are 50% correct, i.e., the firm knows nothing about the government's decision and is merely guessing ($z_O = z_N = 0.5$), and (iii) observations which are 90% correct because, for example, the government may have hinted at its intentions regarding closure ($z_O = z_N = 0.9$). Table 2 shows the firm's discounted expected revenues for these three cases. We observe that the firm's revenues are unchanged when the probability of a correct tax status observation changes from 50% to 90%, with a slight difference compared to the 100% (prefect observations) case. This seems counterintuitive, because one would expect the firm to be able to use any information on its true state to its advantage. The explanation, however, can be found in the firm's optimal policy which is constant and identical for all cases in Table 2, namely, that the firm's best action in the beginning of each fiscal year is to apply for the closure option ($v = 1$) and declare to the government as little profit as possible ($u = 1$) regardless of its belief of being in an O_i vs an N_i state.

Previous work [11] has shown that the tax parameters in the range used in Greece in recent years are such that they encourage tax evasion. What we find here is that the incentive they create for tax evasion is such that the firm's uncertainty as to its state is unimportant because the optimal policy—even in the face of that uncertainty—is to always "cheat." However, this might change—and should, from a policy viewpoint—for other combinations of tax parameters (e.g., higher penalties). Numerical experimentation shows that there is a range of values for r, β, and ℓ where state observation probabilities do become important with regard to the firm's optimal reward and policy. For example, for a tax penalty of $\beta = 9$ (that is, 9 times any tax that went unpaid due to tax evasion) and closure cost $\ell = 0.04$ (or 4% of the firm's profit) we notice a difference in expected rewards between the cases of 50%

Table 2 Comparison of the firm's expected reward when the closure option is offered with a probability of $q = 0.2$

Prob. of correct observation	100%	50%	90%
Exp. reward	3307.1	3309.7	3309.7

The experiments were run with $r = 0.24$, $\beta = 0.24$, $\ell = 0.023$, and a 5% overall audit probability. Numbers are expressed in % of the firm's annual profit, discounted at a 3% annual rate of inflation

Table 3 Comparison of the firm's expected rewards with 50% vs. 90% probability of correctly observing its tax status when the closure option is available with probability $q = 0.2$

Prob. of correct observation	50%	90%
Exp. reward	2630.6	2648

The tax parameters were set to $r = 0.24$, $\beta = 9$, $\ell = 0.04$, and a 5% overall audit probability. Rewards are expressed in % of the firm's annual profit, discounted at a 3% annual rate of inflation

and 90% probability of correct tax status observation, as shown in Table 3 where the firm obtains higher rewards through tax evasion when more certain of its tax status.

The difference in discounted expected rewards for the two cases of Table 3 varies for other combinations of tax parameters, and—although it is beyond the scope of this paper—it would be of interest to "map" the (r, β, ℓ) space in order to quantify the increase in expected reward that the firm could obtain as a function of the "amount of uncertainty" in its state observations.

With respect to the closure option (and other such amnesties), our results suggest that unless the "surrounding" tax environment is sufficiently strict in terms of penalties and cost of the option, the latter should not be offered because it helps the firm achieve a higher expected reward through tax evasion (which means that tax revenues are reduced). If the tax parameters are appropriately set, then the closure option could be useful if taxpayers can be kept from knowing about it in advance.

3.3 The Role of Statute of Limitations

Our prototypical tax system has thus far included a 5 year "window" within which the government is allowed to audit the firm's tax statements. This has been the legal window in Greece, for example. However, the effect of extending this statute of limitations has not been investigated. To quantify the effect of such an extension, we considered the set of audit probability p and tax penalty β combinations, and determined those values for which the firm's optimal policy is to behave honestly (i.e., use $u = 0$ in every state). As a result of the firm's risk-neutrality (and consequent linearity of the reward function) there exists an "honesty boundary" in the form of a curve in the (p, β) space, above which (high penalties) the firm's decision is to never conceal any profit, and below which (lower penalties) the firm conceals its profits in at least one state.

Any point on the boundary can be computed relatively easily by fixing p and using bisection on β, each time calculating the firm's optimal policy and checking whether it is completely "honest." For the case where there is no uncertainty in state observations this can be done via simple value iteration. We thus modified our model to increase the statute of limitations, from the nominal $L = 5$ to $L = 6, \ldots, 10$ years, and computed the corresponding honesty boundaries for comparison with the 5-year case. The various boundaries are shown in Figure 3 for the case where

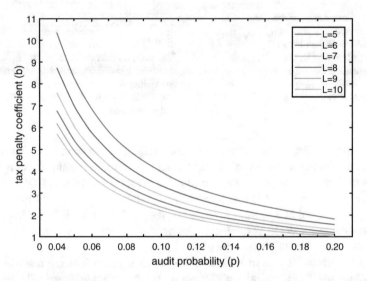

Fig. 3 Firm's "honesty boundaries" in the p-β space when the closure option is never available, as the statute of limitations on tax audits increases from 5 to 10 years. Above each curve the firm's optimal policy is $u_k = 0$, i.e., to always declare all profit

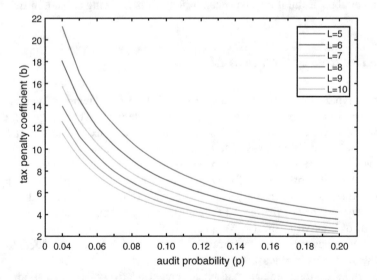

Fig. 4 Firm's "honesty boundaries" in the p-β space when the closure option is available with probability $1/L$ each year, as the statute of limitations L on tax audits increases from 5 to 10 years. Above each curve the firm's optimal policy is $u_k = 0$, i.e., to always declare all profit

Table 4 Average percentage reduction in the tax penalty β which is necessary to eliminate tax evasion, as the statute of limitations on audits, L, increases

Year-to-year/closure prob	q=0	q=1/L
5–6	15.67%	15.10%
6–7	12.87%	12.88%
7–8	10.61%	11.7%
8–9	8.91%	10.28%
9–10	7.63%	8.85%
Overall 5–10	44.74%	46.6%

the closure option is never available, and in Figure 4 for closure available with a fixed nonzero probability. As the time horizon on audits increased, we adjusted the probability of the closure option being offered to $q = 1/L$ so that the firm has, on average, one chance at closure within the statute of limitations period, regardless of that period's length. The first thing to note is that as the audit probability increases there is a lower tax penalty required to enforce complete honesty, and that the marginal effect is most pronounced for low p. Furthermore, as the statute of limitations increases, the entire honesty boundary moves downward, so that again lower penalties are required to remove the incentive for tax evasion. This is to be expected, because the government is given more opportunities to catch the firm at cheating and thus the firm behaves more honestly. Thus, from a technical point of view, the government could extend the auditing period and be able to use more reasonable penalties to deter tax evasion, although unless the audit rates are also increased, those penalties will still have to be very high (a factor of almost 10 for a 5% audit probability even for $L = 10$). In Figure 3, where there is no probability of amnesty, the honesty "threshold" is always lower—and involves less exorbitant tax penalties—than that in Figure 4.

To quantify the reduction of the honesty threshold for β as L varies, we calculated the threshold's year-to-year average percent variation, and also the total variation between the 5- and 10-year cases. The results are shown in Table 4. We note that each yearly extension of the statute of limitations reduces the average tax penalty coefficient required to make tax evasion unprofitable by a significant margin. Combined with our previous results, this points towards (i) a long statute of limitations, (ii) an increased audit probability, and (iii) avoidance of tax amnesties, as useful policy tools.

4 Conclusions

Motivated by the problem of tax evasion and the need for computational tools that may be used to elucidate the behavior of tax entities, we have described a POMDP which models the behavior of a self-interested risk-neutral firm in a tax system which includes random audits, penalties, and occasional amnesties. Using our model, together with a point-based approximation technique, we were able to compute the firm's optimal decisions regarding whether or not to apply for amnesty (in case it is given) and how much of its profit to conceal from authorities.

Our model is more realistic than previously proposed Markov-based models of firm behavior [11, 12] in that the firm is uncertain as to its tax status and must file its taxes before it knows whether its latest filing will be subject to amnesty or not. Using the Greek tax system as a case study, we confirmed previous results suggesting that for range of tax parameters commonly in use there, the firm's optimal policy is to always opt for amnesty and conceal as much of its profit as possible. This is a consequence of the firm's risk neutrality, and the fact that the audit probability and tax penalty coefficients are too low to be effective.

With respect to the role of the firm's conditional observations, we saw that if the tax parameters do not have sufficient deterrent value, the firm's uncertainty as to its true tax status is unimportant. Among other things, this means that the government need not worry about information "leaks" regarding the tax amnesty which, however, is rendered less effective. If tax revenues are to increase (in part by keeping firms in the dark about an upcoming amnesty), some combination of the tax penalties, audit rates, and cost of amnesty must rise significantly from the levels studied here.

Finally, we identified the set of tax rates and tax penalties which eliminate tax evasion in our model. The resulting curves in the tax rate—tax penalty space, show, among other things, how frequent tax audits must be in order for the government to be able to discourage tax evasion using realistic tax penalties. We also found that the "border" between tax evasion and honest behavior shifts significantly towards lower penalties as the statute of limitations on tax audits is extended.

Opportunities for future work include a computational study to explore the space of tax rates, penalty coefficients [31], and audit rates in order to find the settings in which the firm's uncertainty as to its true state has the greatest impact on its optimal long-term revenues. From a policy viewpoint, this would suggest parameter values for which government revenues increase as long as the government can keep any upcoming amnesty a surprise. Finally, it would be interesting to extend the model presented here by considering a risk-averse firm and explore ways of solving resulting POMDP in the presence of the nonlinearity that risk-aversion introduces in the reward function, in the spirit of [12].

Appendix

Transition Matrix when Firm Applies for Closure

Markov transition matrix for the case where the firm asks to use the closure option. The statute of limitations is assumed to be $L = 5$ years

$$T_1 = \begin{bmatrix}
p_V & p_V & p_V & p_V & p_V & p_O & p_O & p_O & p_O & p_O & 0 & 0 & 0 & 0 & 0 \\
0 & 0 & 0 & 0 & 0 & 0 & 0 & 0 & 0 & 0 & p_N & 0 & 0 & 0 & 0 \\
0 & 0 & 0 & 0 & 0 & 0 & 0 & 0 & 0 & 0 & 0 & p_N & 0 & 0 & 0 \\
0 & 0 & 0 & 0 & 0 & 0 & 0 & 0 & 0 & 0 & 0 & 0 & p_N & 0 & 0 \\
0 & 0 & 0 & 0 & 0 & 0 & 0 & 0 & 0 & 0 & 0 & 0 & 0 & p_{Nf} & p_{Nf} \\
q & q & q & q & q & q & q & q & q & q & 0 & 0 & 0 & 0 & 0 \\
0 & 0 & 0 & 0 & 0 & 0 & 0 & 0 & 0 & 0 & q & 0 & 0 & 0 & 0 \\
0 & 0 & 0 & 0 & 0 & 0 & 0 & 0 & 0 & 0 & 0 & q & 0 & 0 & 0 \\
0 & 0 & 0 & 0 & 0 & 0 & 0 & 0 & 0 & 0 & 0 & 0 & q & 0 & 0 \\
0 & 0 & 0 & 0 & 0 & 0 & 0 & 0 & 0 & 0 & 0 & 0 & 0 & q & q \\
p_{lV} & p_{lV} & p_{lV} & p_{lV} & p_{lV} & p_{lO} & p_{lO} & p_{lO} & p_{lO} & p_{lO} & 0 & 0 & 0 & 0 & 0 \\
0 & 0 & 0 & 0 & 0 & 0 & 0 & 0 & 0 & 0 & p_{lN} & 0 & 0 & 0 & 0 \\
0 & 0 & 0 & 0 & 0 & 0 & 0 & 0 & 0 & 0 & 0 & p_{lN} & 0 & 0 & 0 \\
0 & 0 & 0 & 0 & 0 & 0 & 0 & 0 & 0 & 0 & 0 & 0 & p_{lN} & 0 & 0 \\
0 & 0 & 0 & 0 & 0 & 0 & 0 & 0 & 0 & 0 & 0 & 0 & 0 & p_{lNf} & p_{lNf}
\end{bmatrix}.$$

Here, p is the fraction of tax filings which the government can audit each year regardless of firm status. For our purposes p is nominally set to 0.05 (5%). A 20% of those audits (1%) are spread over firms that have been unaudited for up to 3 years, and the remaining 80% of audits (nominally 4%) are for those that have gone unaudited for 4 or more years; q is the probability of the government offering the closure option; $p_V = (p - 0.8p)/4$ is the probability of a repeat audit (i.e., for the second year in a row); $p_O = (p/5)/4$ is the probability of being audited after using closure; $p_N = (p/5)/4$ is the probability of an audit if the firm has been unaudited for 1–3 years; $p_{N_f} = (p/5)4$ is the probability of being audited if the firm has been unaudited for 4 or more years (and thus some of its tax statements are about to pass beyond the statute of limitations); and $p_{lV} = 1 - q - p_V$, $p_{lO} = 1 - q - p_O$, $p_{lN} = 1 - q - p_N$, and $p_{lNf} = 1 - q - p_{Nf}$.

Transition Matrix when Firm Declines Closure

Markov transition matrix for the case where the firm forgoes the closure option. The statute of limitations is assumed to be $L = 5$ years

$$
T_2 =
\begin{bmatrix}
p_V & p_V & p_V & p_V & p_V & p_O & p_O & p_O & p_O & p_O & 0 & 0 & 0 & 0 & 0 \\
0 & 0 & 0 & 0 & 0 & 0 & 0 & 0 & 0 & 0 & p_N & 0 & 0 & 0 & 0 \\
0 & 0 & 0 & 0 & 0 & 0 & 0 & 0 & 0 & 0 & 0 & p_N & 0 & 0 & 0 \\
0 & 0 & 0 & 0 & 0 & 0 & 0 & 0 & 0 & 0 & 0 & 0 & p_N & 0 & 0 \\
0 & 0 & 0 & 0 & 0 & 0 & 0 & 0 & 0 & 0 & 0 & 0 & 0 & p_{Nf} & p_{Nf} \\
0 & 0 & 0 & 0 & 0 & 0 & 0 & 0 & 0 & 0 & 0 & 0 & 0 & 0 & 0 \\
0 & 0 & 0 & 0 & 0 & 0 & 0 & 0 & 0 & 0 & 0 & 0 & 0 & 0 & 0 \\
0 & 0 & 0 & 0 & 0 & 0 & 0 & 0 & 0 & 0 & 0 & 0 & 0 & 0 & 0 \\
0 & 0 & 0 & 0 & 0 & 0 & 0 & 0 & 0 & 0 & 0 & 0 & 0 & 0 & 0 \\
0 & 0 & 0 & 0 & 0 & 0 & 0 & 0 & 0 & 0 & 0 & 0 & 0 & 0 & 0 \\
p_{lv} & p_{lv} & p_{lv} & p_{lv} & p_{lv} & p_{lo} & p_{lo} & p_{lo} & p_{lo} & p_{lo} & 0 & 0 & 0 & 0 & 0 \\
0 & 0 & 0 & 0 & 0 & 0 & 0 & 0 & 0 & 0 & p_{lN} & 0 & 0 & 0 & 0 \\
0 & 0 & 0 & 0 & 0 & 0 & 0 & 0 & 0 & 0 & 0 & p_{lN} & 0 & 0 & 0 \\
0 & 0 & 0 & 0 & 0 & 0 & 0 & 0 & 0 & 0 & 0 & 0 & p_{lN} & 0 & 0 \\
0 & 0 & 0 & 0 & 0 & 0 & 0 & 0 & 0 & 0 & 0 & 0 & 0 & p_{lNf} & p_{lNf}
\end{bmatrix}.
$$

Again, p is the fraction of tax filings which the government can audit each year regardless of firm status; $p_V = (p - 0.8p)/4$ is the probability of a repeat audit (i.e., for the second year in a row); $p_O = 3[(p/5)/4]$ is the probability of being audited after using closure; $p_N = (p/5)/4$ is the probability of an audit if the firm has been unaudited for 1–3 years; $p_{Nf} = (p/5)4$ is the probability of being audited if the firm has been unaudited for 4 or more years (and thus some of its tax statements are about to pass beyond the statute of limitations); and $p_{lV} = 1 - p_V$, $p_{lO} = 1 - p_O$, $p_{lN} = 1 - p_N$, and $p_{lNf} = 1 - p_{Nf}$.

References

1. Allingham, P., Sandmo, H.: Income tax evasion: a theoretical analysis. J. Public Econ. **1**(6), 988–1001 (1972)
2. Alm, J., Beck, W.: Tax amnesties and tax revenues. Pub. Fin. Rev. **18**(4), 433–453 (1990)
3. Alm, J., Rath, D.M.: Tax policy analysis: the introduction of a Russian Tax Amnesty. Technical Report, Georgia State University, Andrew Young School of Policy Studies (1998)
4. Baldry, J.C.: Tax evasion and labour supply. Econ. Lett. **3**(1), 53–56 (1979). ISSN: 0165-1765
5. Bloomquist, K.M., Koehler, M.: A large-scale agent-based model of taxpayer reporting compliance. J. Artif. Soc. Soc. Simul. **18**(2), 20 (2015)
6. Chari, V.V., Nicolini, J.P., Teles, P.: Optimal Capital Taxation, University of Minnesota and Federal Reserve Bank of Minneapolis, Federal Reserve Bank of Minneapolis- Universidad Di Tella, Universidad Autonoma de Barcelona, Banco de Portugal, Catolica Lisbon SBE, CEPR (2016)
7. Das-Gupta, A., Mookherjee, D.: Tax amnesties in India: an empirical evaluation. World Dev. **23**(12), 2051–2064 (1995)
8. Dutech, A., Scherrer, B.: Partially Observable Markov Decision Processes. Wiley, Hoboken (2013)

9. Gao, S., Xu, D.: Conceptual modeling and development of an intelligent agent-assisted decision support system for anti-money laundering. Exp. Syst. Appl. **36**(2), 1493–1504 (2009)
10. Garrido, N., Mittone, L.: Tax evasion behavior using finite automata: experiments in Chile and Italy. Exp. Syst. Appl. **39**(5), 5584–4492 (2012)
11. Goumagias, N.D., Hristu-Varsakelis, D., Saraidaris, A.: A decision support model for a tax revenue collection in Greece. Decis. Support. Syst. **53**(1), 76–96 (2012)
12. Goumagias, N.D., Hristu-Varsakelis, D., Assael, J.: Using deep Q-learning to understand the tax evasion behavior of risk-averse firms. Exp. Syst. Appl. **101**, 258–270 (2018)
13. Greek Ministry of Finance, Law N.3259/2004 (pol.1034/2005) (in Greek) (2004)
14. Greek Ministry of Finance, Law N.3697/2008 (pol.1130/2008) (in Greek) (2008)
15. Greek Ministry of Finance, Law N.4337/2015 (pol.4337/2015) (in Greek) (2015)
16. Kaplow, R.: Point-based POMDP solvers: survey and comparative analysis. Master's thesis, McGill University (2010)
17. Krishnamurthy, V.: Partially Observed Markov Decision Processes: From Filtering to Controlled Sensing. Cambridge University Press, Cambridge (2016)
18. Kydland, F.E., Prescott, E.C.: Dynamic optimal taxation, rational expectations and optimal control. J. Econ. Dyn. Control **2**(1), 79–91 (1980)
19. Levagi, R., Menoncin, F.: Optimal dynamic tax evasion: a portfolio approach. J. Econ. Dyn. Control **37**(11), 2157–2167 (2011)
20. Liu, A.A.: Tax evasion and optimal environmental taxes. J. Environ. Econ. Manage. **66**(3), 656–670 (2013)
21. Martinez-Vazquez, J., Rider, M.: Multiple modes of tax evasion: theory and evidence. Natl. Tax J. **58**(1), 51–76 (2005)
22. Pineau, J., Gordon, G., Thrun, S.: Point-based value iteration: an anytime algorithm for POMDPs. In: Proceedings of International Joint Conference on Artificial Intelligence, Acapulco, Mexico (2003)
23. Pineau, J., Gordon, G., Thrun, S.: Anytime point-based approximations for large POMDPs. J. Artif. Intell. Res. **27**, 335–380 (2006)
24. Poupart, P.: Exploiting structure to efficiently solve large scale partially observable Markov decision processes. Ph.D. Thesis, Department of Computer Science, University of Toronto (2005)
25. Ross, J.M., Buckwalter, N.D.: Strategic tax planning for state tax amnesties evidence from eligibility period restrictions. Public Finance Rev. **4**(3), 275–301 (2013)
26. Shani, G., Pineau, J., Kaplow, R.: A survey of point-based POMDP solvers. Auton. Agent Multi-Agent Syst. **27**(1), 1–51 (2012)
27. Sondik, E.J.: The optimal control of partially observable Markov decision processes. Ph.D. Thesis, Stanford University (1971)
28. Spaan, M.T.J.: POMDP solving software - perseus randomized point-based algorithm. Universiteit van Amsterdam (2004)
29. Spaan, M.T.J., Vlassis, N.: Perseus: randomized point-based value iteration for POMDPs. J. Artif. Intell. Res. **24**, 195–220 (2005)
30. Tatsos, N.: Economic Fraud and Tax Evasion in Greece (in Greek). Papazisis Publishings, Athens (2001)
31. Vasin, A., Vasina, P.: Tax optimization under tax evasion, the role of penalty constraints. Econ. Educ. Res. Consortium **1**, 1–44 (2009)
32. Vlassis, N., Spaan, M.T.J.: A fast point-based algorithm for POMDPs. In: Benelearn 2004: Proceedings of Annual Machine Learning Conference of Belgium and the Netherlands, pp. 170–176, (2004). Also presented at the NIPS 16 workshop 'Planning for the Real-World', Whistler, Canada, Dec. 2003
33. Yitzhaki, S.: Income tax evasion: a theoretical analysis. J. Public Econ. **3**(2), 201–202 (1974)

Printed in the United States
By Bookmasters